AI

for
Finance Professionals

Prabash Galagedara

First published by Busybird Publishing 2024

ISBN:
Paperback: 978-1-923216-00-6
Ebook: 978-1-923216-01-3

Cover image: Adobe Stock

Cover design: Busybird Publishing

Layout and typesetting: Busybird Publishing

Busybird Publishing
2/118 Para Road
Montmorency, Victoria
Australia 3094
www.busybird.com.au

*You must learn about AI
before AI learns about you!*

In you, I see a mirror of my own childhood –
threads of self-doubt, stubbornness, dreams,
adventure, and unbridled joy.

Kevon, my dear son, you are the self-claimed editor of
my work, a role you embrace with enthusiasm that
brightens every word I write. With each page you turn
and every word you read, remember this: you are the
heartbeat behind every chapter of my work.
This journey of words is not solely mine; it is ours,
a shared odyssey, penned with all the love and pride
that swells in my heart for you.

May this book be more than just a collection of pages
and stories. Let it be a guide, a steadfast companion on
the journey of many children around the world, and a
tangible reminder of how words can empower
a generation of youth to achieve their potential.

Testimonials

In the nexus of artificial intelligence (AI) and human progress, Prabash Galagedara's *AI for Finance Professionals* emerges as a beacon of understanding and guidance. With a profound grasp on the requirements and aspirations of finance professionals worldwide, Prabash, a valued member of the Institute of Chartered Accountants of Sri Lanka, has crafted a work that not only illuminates the path forward but also emboldens us to step into a future where technology and human insight converge to redefine the finance profession. This book is a testament to the author's visionary approach and his commitment to leveraging AI as a force for good, enhancing decision-making, efficiency, and innovation across the financial landscape.

As we navigate this pivotal era, where the potential of AI unfolds to reshape our professional and personal lives, his contributions stand out as both a roadmap and a source of inspiration. His work underscores the transformative power of AI, while also attending to the ethical, practical, and strategic dimensions of its application in finance. It is with great admiration and respect that I extend my best wishes to Prabash in his endeavours to make the world a better, more insightful place through the thoughtful integration of AI into finance. May his efforts inspire countless professionals, including fellow members of the Institute of Chartered Accountants of Sri Lanka and members of the Professional Accountancy Organisations in the region, to explore, adapt, and advance the frontiers of AI, driving progress and prosperity for all.

Heshana Kuruppu

President of the South Asian Federation of Accountants & President of the
Institute of Chartered Accountants of Sri Lanka

•

An immensely valuable and practical guide for students and professionals in the field of finance globally.

Martin ABC Hansen

Global Data Leader and Managing Director of Copenhagen Data

•

This book brilliantly demystifies complex AI concepts, making them accessible to finance experts and novices alike.

Andrew Warren

Global Leader in Responsible AI

•

AI for Finance Professionals stands out as a critical resource for those at the forefront of integrating AI in business. The author's expertise on the subject makes complex concepts accessible and practical. Whether you are a seasoned finance expert or new to the field of AI, the insights contained in this book will equip you with the knowledge and foresight to lead in the age of AI.

Dr Gladys Moran Paredes

Chancellor & CEO of Universidad Maria Auxiliadora, Lima, Peru

•

AI for Finance Professionals is a ground-breaking book that is going to shape the future of finance and accountancy education. This is a must-read for both students and experienced professionals. I really appreciate the author's dedication and commitment to this extraordinary and proactive project in contributing to new knowledge in AI.

Prof. Prem Yapa

RMIT University, College of Business and Law

•

This book is an invaluable resource for those interested in the intersection of finance, AI, and technology.

Kristen Whittle

Architect, AI Systems Innovator, Urban Designer

•

Depth and practicality underpin this exploration of AI within the context of finance with an emphasis on framework-driven processes and prioritising the understanding of foundational AI principles over mere tool mastery. This book is a valuable resource enabling readers to gain insights into deploying AI solutions strategically, ensuring alignment with organisational objectives and regulatory standards, thereby facilitating informed decision-making and fostering innovation within the finance industry.

Jannat Maqbool

AI Enthusiast & Co-founder of CPAI Australia

•

This book implores all finance professionals to learn about AI before AI learns about finance! There is no better way of ensuring this than by studying the many finance-AI interfaces explored in this book using the case-study approach. A must-read for all finance professionals who want to be ahead of the curve.

Prof. Janek Ratnatunga

CEO of The Institute of Certified Management Accountants,
Australia & New Zealand

Contents

1

Introduction

During the advent of the artificial intelligence era, finance professionals, from graduates to seasoned members, actively participated in reshaping the field. Global professional bodies collaborated with startups, the Big 4, Big Tech, and emerging talent to foster a cohesive ecosystem which encompasses curriculum updates, technological advancements, and talent cultivation. This comprehensive blueprint sets a precedent, now increasingly emulated by other professional bodies, marking a new era of interdisciplinary growth and innovation.

It is impressive how the finance profession is quickly embracing innovation in this AI era to revolutionise best practices and transform the industry. The term finance professional includes but is not limited to, chief financial officers, financial accountants, management accountants, tax professionals, auditors, and forensic accountants. The profession is one of the oldest and most sought-after for centuries and has gone through many revolutionary changes in the business and industry environment.

The origins of the finance profession can be traced back to ancient civilisations. However, the Renaissance marked a significant leap in accounting with the development of double-entry bookkeeping by Italian mathematician Luca Pacioli. In his seminal work *Summa de Arithmetica*, published in 1494, Pacioli outlined the principles of double-entry accounting, a system where each transaction has equal and opposite effects on at least one account. This system provided a more accurate way to track financial transactions and became the foundation of modern accounting.

The Industrial Revolution brought about significant changes in business structures and operations. The company structure became the most admired corporate structure which, even to date, is the most common and dominant business structure across the globe. As enterprises grew larger and more complex, the demand for accurate financial information increased. The 19th century witnessed the establishment of professional accounting bodies, such as the Institute of Chartered Accountants, marking the formalisation of the accounting profession.

The 20th century saw the expansion of corporate entities and the emergence of stock exchanges. This necessitated more rigorous and standardised financial reporting. Governments worldwide introduced regulations to ensure transparency and protect investors. The regulation of stock exchanges laid the groundwork for financial disclosure requirements, which cemented the role of qualified accountants in companies worldwide.

Advancements in technology revolutionised finance practices. The introduction of computers in the latter half of the 20th century streamlined bookkeeping processes. The manual bookkeeping on ledgers gradually disappeared. Spreadsheets, accounting software, and Enterprise Resource Planning (ERP) systems enhanced accuracy and efficiency in financial management.

Globalisation necessitated a move towards harmonising accounting standards. The International Financial Reporting Standards (IFRS) emerged as a set of international accounting standards aimed at creating consistency in financial reporting across countries and industries. This made comparing the records of entities regardless of their geographical location easy.

The 21st century brought new challenges and opportunities for finance professionals. Increasingly complex business structures, cybersecurity threats, and the rise of cryptocurrencies are among the issues accountants must navigate. Moreover, the profession is embracing data analytics, AI, and machine learning to derive insights from vast amounts of financial data.

In short, the finance profession has evolved from simple record-keeping in the past to a sophisticated discipline central to modern business operations. The profession continues to adapt to technological advancements, regulatory changes, and global economic shifts, maintaining its crucial role in financial management and decision-making.

The future of the profession and its potential rests with today's accountants – students, practitioners, administrators, and regulators.

1. Embracing technology

Today's finance professionals must be proficient in leveraging technology to enhance efficiency and accuracy. Cloud-based accounting systems, data analytics, AI, and automation are becoming integral to the profession. Accountants should be well-versed in these tools to streamline processes, analyse large volumes of data, and provide valuable insights for decision-making.

The integration of AI in the field of finance has indeed sparked discussions and concerns about the potential impact on jobs within the profession. While AI presents opportunities for increased efficiency, accuracy, and advanced data analysis, it also raises questions about the future role of accountants akin to the emergence of accounting software in the late 20th century.

2. Data analytics and business intelligence

The increasing volume of data available presents both a challenge and an opportunity for finance professionals. Analysing financial and non-financial data can uncover valuable insights for strategic decision-making.

Accountants should develop skills in data analytics and business intelligence to extract meaningful information and contribute to the organisation's overall performance. This doesn't mean every accountant has to be a skilled programmer. However, accountants must understand the basic principles of data science and the use of AI to make them more effective and efficient in their field.

3. Strategic business partner

The role of the finance professional is evolving from a traditional number-cruncher to a strategic business partner. Accountants should actively participate in strategic planning, offering financial insights and recommendations to support achieving goals. This involves understanding the broader business context, industry trends, and potential risks and opportunities. About a third of Fortune 500 CEOs spent the first few years of their careers developing a strong foundation in finance. Most of them are even qualified accountants who sharpened their business skills as strategic finance partners early in their careers.

4. Cybersecurity and risk management

With the increasing reliance on digital systems, finance professionals need to be well-versed in cybersecurity and risk management. Protecting sensitive financial information from cyber threats and ensuring compliance with data protection regulations are critical aspects of the modern accountant's role.

The auditors have a duty of care to understand the overall risk profile of the organisation and mitigating activities before forming an audit opinion. The financial impact of the Equifax cyber-attack in 2017, the identity theft of over 160 million individuals, led to immediate costs, long-term repercussions, and ongoing investments in cybersecurity and reputation management. The stolen identities resulted in scams and fraud on individuals who still to date painfully deal with financial losses. The incident profoundly affected Equifax's financial standing, emphasising the broader consequences organisations face when entrusted with securing sensitive consumer data.

5. Environmental, social, and governance

Net-zero commitment is one of the top global priorities and commitments for the next few decades. Climate stakeholders are increasingly interested in a company's impact on the environment, social, and governance (ESG). Accountants should be familiar with ESG reporting standards and play a role in ensuring transparent and accurate disclosure of non-financial information.

6. Continuous professional development

The pace of change in the business environment demands accountants engage in continuous learning. Staying updated on new accounting standards, regulations, and emerging technologies is crucial for professional development. Accountants should actively seek opportunities for upskilling and adapting to industry changes.

7. Ethical leadership

Ethical considerations have always been central to finance, but they are gaining even more prominence in today's business world. Accountants must demonstrate ethical leadership by adhering to a strong code of conduct, promoting transparency, and addressing ethical dilemmas proactively.

8. Cross-functional collaboration

Collaboration across departments is increasingly important. Finance professionals should work closely with colleagues in finance, operations, marketing, and other areas to gain a holistic understanding of the business. This collaborative approach enhances the accountant's ability to provide well-informed financial advice.

9. Globalisation and international standards

Finance professionals operating globally need to be familiar with international accounting and auditing standards. With businesses expanding across borders, understanding and complying with global financial reporting standards is essential for accurate and comparable financial information.

10. Adaptability and agility

The ability to adapt and embrace change is crucial for accountants. Whether it is responding to regulatory changes, technological advancements, or shifts in market dynamics, finance professionals should be agile and able to pivot their strategies accordingly. Professional bodies play an important role in preparing both students and members to face challenges and capitalise on opportunities but the mindset of those who are in the field is more important than ever.

The finance professional of the future needs to be a dynamic, tech-savvy, and strategically minded professional. By embracing technological innovations, staying informed about industry trends, and actively contributing to strategic decision-making, today's accountants can position themselves as indispensable assets in an ever-evolving business landscape. Accountants must work hard to maintain this position for the foreseeable future.

Finance professionals must learn about AI before AI learns about finance!

2

Innovation

Humans are an inventive species that have exerted a profound influence on both their own existence and the surrounding environment for millennia. From the Stone Age to the modern world, individuals have continuously invented new tools and expanded their knowledge, contributing to the world we inhabit today.

Innovation Theory

There are so many scholarly and seminal works about innovation but none other than the late Harvard Business School Professor Clayton Christensen's Disruptive Innovation Theory has had a profound impact on the 21st century. Christensen's theory is considered by some the most influential business idea of the 21st century, and he is the most influential management thinker of our time.

Theories of innovation attempt to answer many questions. However, the fundamental question the theories intend to answer is "Why do most successful and large ventures, corporations, nations, and sports fail in the long term?" There are so many examples in recent history: Blockbuster vs. Netflix, Kodak vs. digital photography, and taxis vs. Uber and Lyft.

Inspired by Christensen's game-changing model, I developed the following for this context. The framework identifies two types of innovation: continuing innovation and game-changing innovation. The innovation process usually happens over several steps:

1. Existing (established) successful businesses focus on continuing innovation by improving their products or services to appeal to their most profitable customers.

2. This continuing innovation ignores a large section of customers who have no other options in the market.

3. Entrants target this ignored market segment and gain traction by meeting their needs at an affordable cost compared to what is offered by existing businesses.

4. Entrants provide a new game-changing innovative product solution, often with a new business model to customers.

5. Existing businesses ignore the new entrant, continuing to focus on their more profitable customers.

6. Entrants eventually improve their products and services by offering solutions that appeal to the incumbent's businesses' customers.

7. Once the new entrant has begun to attract these customers at scale, game-changing innovation has occurred.

Continuing Innovation Model

Continuing innovation is making existing products and services better for consumers. This is a crucial aspect of organisational growth and longevity, involving the ongoing improvement and evolution of products, processes, or services within an existing market or industry.

Unlike game-changing innovation, which often introduces entirely new paradigms, continuing innovation seeks to build upon and enhance existing products or practices. An organisation's processes are typically geared towards continuing innovation, making existing products and services better with more features customers often don't need. Below are key aspects about continuing innovation.

1. Incremental improvements

Continuing innovation involves ongoing efforts to refine and enhance existing products or processes. This can include incremental improvements, efficiency gains, or the addition of new features based on customer feedback and market trends. Capital is allocated to support business cases that generate more benefits and higher returns on investments. Managers in organisations focus heavily on targeting existing customers with better products as their incentives are often based on revenue and cost targets.

2. Ask customers what they want

Understanding and responding to customer needs are central to continuing innovation. By actively seeking and incorporating customer feedback, organisations can tailor their offerings to meet evolving preferences, ensuring continued relevance in the market.

3. Quality improvements

Organisations focus on maintaining or improving the quality of their products or services. This not only sustains customer satisfaction but also reinforces the brand's reputation for reliability and excellence.

4. Cost productivity

Continuing innovation often includes efforts to optimise processes, reduce costs, and improve operational efficiency. This can lead to increased profitability while maintaining competitive pricing in the market.

5. Market expansion

Continuing innovation can involve exploring new markets or segments for existing products. This may include adapting products for different customer segments or geographic regions.

6. Technology integration

Embracing and integrating new technologies is a common strategy for continuing innovation. This can involve incorporating the latest advancements to enhance product features, performance, or user experience.

7. Employee involvement

Fostering a culture of innovation within the organisation is essential. Encouraging employees at all levels to contribute ideas, experiment with new approaches, and participate in continuous improvement initiatives strengthens the capacity for continuing innovation.

Innovation is about the ongoing refinement, enhancement, and adaptation of products or processes within existing markets. It is a dynamic process that requires a customer-centric mindset, a commitment to quality, and a willingness to embrace technological advancements. By fostering a culture of continuous improvement, organisations can ensure their relevance, competitiveness, and resilience in an ever-changing business landscape.

Game-changing Innovation Model

Game-changing innovation, in contrast, doesn't attempt to improve products to existing customers in existing markets. Rather, they disrupt and redefine the existing market trends by introducing products and services that are not as good as what is currently available for consumers. However, game-changing products and services offer other benefits such as simplicity, convenience, and lower prices.

Game-changing innovation is a transformative force that introduces groundbreaking ideas, products, or services, often reshaping entire industries and challenging established norms. Unlike continuing innovation, which focuses on incremental improvements within existing frameworks, game-changing innovation brings about radical change, creating new markets or fundamentally altering existing ones.

Below are key characteristics of game-changing innovation.

1. New market opportunity

Game-changing innovations often give rise to entirely new markets or niches that were not previously recognised or served. These innovations can tap into unmet needs or create demand that did not exist before.

2. Technology-driven

The introduction of new technologies or novel applications of existing ones can lead to groundbreaking solutions that redefine the possibilities within a given industry.

3. Customer focus

Successful game-changing innovations often stem from a deep understanding of user needs and a commitment to addressing pain points or inefficiencies in a way that significantly improves the user experience. To paraphrase Christensen from his article titled "Know Your Customers' 'Jobs to Be Done'", "People don't buy products; they 'hire' them to do jobs, such as solving a problem or fulfilling a desire."

4. Challenges status quo

Game-changing innovation challenges established norms and market leaders. It can displace existing products or services, even if they have been successful for an extended period, by offering more efficient or cost-effective alternatives.

5. Uncertainty and risk

Game-changing innovation inherently involves a degree of risk and uncertainty. It may not be immediately embraced by the mainstream market, and the path to success may involve overcoming resistance or scepticism.

6. *Adoption curve*

Game-changing innovations typically follow an adoption curve, taking time to gain broader acceptance. The rate of adoption may vary, and successful game-changing innovations can rapidly become part of mainstream culture.

7. *Continuous evolution*

Game-changing innovation is an ongoing process. As technologies advance and market dynamics shift, new disruptions can emerge, requiring organisations to stay agile and adaptable to remain competitive.

Examples of Game-changing Innovation

1. *Smartphones*

Nokia, once a dominant force in the mobile phone industry, provides an interesting case study on game-changing innovation. While Nokia was a market leader for many years, its decline was associated with its inability to adapt to the disruptive changes brought about by the smartphone revolution.

Nokia was a pioneer and market leader in the mobile phone industry, particularly in the early 2000s. Its traditional mobile phones, known for their durability and reliability, dominated the global market. Despite its success, Nokia initially resisted the shift toward smartphones.

The introduction of the iPhone by Apple in 2007 marked a game-changing moment in the mobile phone industry. Its touchscreen interface, robust app ecosystem, and multimedia capabilities redefined user expectations and set a new standard for smartphones.

While traditional mobile phones were Nokia's strength, the company was slow to recognise the potential of touchscreen devices and the emerging app ecosystem, central elements of game-changing innovation and consumer preferences in the mobile phone industry.

Nokia's commitment to its Symbian Operating System became a hindrance.

As competitors embraced more modern and user-friendly operating systems (iOS and Android), Nokia's reliance on Symbian contributed to its decline in the smartphone market. In an effort to recover, Nokia entered into a partnership with Microsoft in 2011 to adopt the Windows Phone operating system. However, this move did not reverse its fortunes, and Nokia continued to lose market share.

The success of smartphones was not just about hardware but also about the app ecosystem. Nokia's Ovi Store, while introduced to compete with app stores, did not gain the same traction as Apple's App Store or Google Play. Nokia's decline in the smartphone market led to the eventual sale of its devices and services division to Microsoft in 2014. This marked the end of Nokia's era as a major player in the mobile phone industry.

Nokia's story underscores the importance of recognising disruptive trends in the technology landscape and adaptability. Companies that fail to innovate and pivot in response to emerging technologies risk becoming obsolete, even if they were once market leaders.

In conclusion, Nokia's experience with game-changing innovation highlights the critical need for established companies to remain vigilant, adaptable, and responsive to changing market dynamics. The smartphone revolution reshaped the industry, and Nokia's inability to effectively navigate this disruptive shift had profound implications for its market position. Today, Apple is one of the largest technology companies in the world, with a market capitalisation of more than 2 trillion US dollars. The case serves as a cautionary tale for companies across industries about the risks of complacency in the face of disruptive technological advancements.

2. Ride-sharing platforms

The taxi industry has experienced significant disruption due to innovative technologies and new business models. The emergence of ride-sharing platforms, such as Uber and Lyft, has transformed the traditional taxi business and challenged its established norms.

Ride-sharing platforms introduced a new and more efficient model for connecting riders with drivers using mobile apps. This innovation disrupted the traditional taxi industry by providing a convenient and often more cost-effective alternative. Ride-sharing platforms leveraged technology to streamline the booking process, payment options, and provide real-time tracking.

The introduction of a peer-to-peer model allowed private individuals to use their personal vehicles as a means of transportation, expanding the pool of available drivers. Ride-sharing platforms prioritised jobs, moving people from Place A to Place B seamlessly. This problem-centric approach attracted users away from the traditional taxi industry, which was a painful experience for riders. Ride-sharing services provide on-demand transportation, reducing wait time and increasing accessibility. Users can request a ride at any time and from almost any location.

The implementation of dynamic pricing models, where fares adjust based on demand, provided transparency and flexibility. Traditional taxis often have fixed pricing, and the introduction of dynamic pricing was a notable shift. The game-changing nature of ride-sharing platforms led to regulatory challenges.

Traditional taxi operators raised concerns about unfair competition, safety standards, and adherence to existing regulations. As ride-sharing gained popularity, traditional taxi services experienced a decline in market share. Consumers increasingly opted for the convenience, affordability, and user-friendly features offered by ride-sharing platforms. Traditional taxi drivers faced economic challenges as they struggled to compete with the lower prices and greater convenience offered by ride-sharing services.

In response to the innovation, some traditional taxi companies adopted mobile apps and improved their technology to compete with ride-sharing platforms. This shift has prompted traditional taxi companies to re-evaluate and adapt their business models to remain relevant.

Despite the disruption and enormous value to consumers, there remain many safety and reliability concerns for consumers. While ride-sharing companies continue to focus on the issues, customer expectations remain to be fully met by the industry.

In summary, game-changing innovation, driven by the introduction of ride-sharing platforms, has reshaped the taxi industry. The shift towards a technology-driven, jobs-centric, and on-demand model has had a profound impact on traditional taxi services, prompting both challenges and adaptations within the industry. The case of the taxi industry illustrates the importance of embracing innovation and adapting to changing consumer preferences in a rapidly evolving business landscape.

3. Streaming services

Blockbuster's demise is often cited as a classic example of a company failing to adapt to game-changing innovation, particularly in the context of the rise of streaming services. Netflix, a game-changing innovation, played a role in the downfall of Blockbuster. Blockbuster was a dominant force in the video rental industry, offering a vast selection of movies and video games for in-store rentals. Its business model relied on customers visiting physical stores to rent and return videos and DVDs. The game-changing innovation in this context was the emergence of digital streaming services led by companies like Netflix. These services allowed users to stream movies and TV shows online, eliminating the need for physical rentals.

Netflix introduced a subscription-based model that provided unlimited streaming for a fixed monthly fee. This offered convenience, cost-effectiveness, and a vast content library without the hassle of travel, time constraints, and fees associated with physical rentals. Users could watch content anytime, anywhere, on a variety of devices.

Blockbuster initially failed to recognise the game-changing potential of streaming. The company did experiment with a DVD-by-mail service, but it did not pivot decisively toward online distribution.

Blockbuster's entry into the digital space came relatively late, with the launch of its own online rental platform. However, by this time, streaming services had already gained significant traction.

Blockbuster struggled to change its business model to compete with subscription-based, on-demand streaming services. Its reliance on in-store rentals became a weakness as consumers increasingly favoured on-demand streaming options. Blockbuster faced financial challenges as its traditional business model became obsolete. The company closed many of its physical stores, unable to sustain the costs associated with maintaining a large retail presence. Blockbuster eventually filed for bankruptcy in 2010. While the brand still exists in some form, it was acquired by DISH Network, and the decline of Blockbuster as a once-dominant player in the video rental industry was evident.

Blockbuster's downfall highlights the importance of adapting to game-changing innovations and evolving consumer preferences. The failure to recognise the shift toward streaming ultimately led to its demise.

Blockbuster's story also serves as a cautionary tale about the consequences of not adapting to game-changing innovations. The rapid rise of streaming services transformed the entertainment landscape. The case underscores the importance of staying attuned to industry trends and embracing innovation to remain relevant in a dynamic business environment.

Types of Game-changing Innovation

While large and incumbent ventures continue to prioritise continuing innovation, the new entrants with game-changing innovations challenge the leader and eventually destroy their positions.

The game-changing innovation has two facets: market-creating disruption and market-altering disruption.

1. Market-creating disruption

Market-creating disruption, also known as new-market innovation, refers to the process of introducing products or services that target an entirely new and often untapped market. This type of game-changing innovation creates a fresh demand by addressing the needs of customers who were not previously considered part of the market.

Innovators identify opportunities to address the needs of non-consumers or excluded segments, effectively expanding the overall market. Instead of competing within existing markets, game-changers focus on serving non-consumers, those who, for various reasons, are not currently benefiting from or participating in the market.

Market-creating game-changing innovation often introduces solutions, products, or services that are simpler, more affordable, or more accessible (at entry level), making them attractive to a broader audience. This allows a wider range of consumers, including those with limited resources or expertise, to participate. The goal is to expand the customer base by reaching individuals or segments that were previously overlooked or excluded. This can lead to rapid market growth and the establishment of a new industry.

The introduction of affordable and featureless mobile phones in emerging markets created new opportunities for connectivity and access to digital services. Microfinance institutions disrupted the traditional banking sector by providing financial services to individuals and businesses in underserved or unbanked regions. Chotu Kool, a portable, compact refrigerator designed for use in rural and off-grid areas was developed by Godrej & Boyce, an Indian conglomerate, and was introduced to address the refrigeration needs of communities with limited access to electricity in India.

Technological advancements often play a crucial role in market-creating game-changing innovation. The use of technology can significantly lower costs, improve efficiency, and enable innovative solutions for new consumers. Market creators may emerge by applying innovations from one industry to another. Cross-industry collaboration and the transfer of ideas can lead to breakthroughs in serving new markets.

Market-creating game-changing innovation is not without challenges. Uncertainties related to consumer adoption, regulatory environments, and competition may need to be navigated. Successful market-creating game-changing innovation can have profound economic and social impacts. It can contribute to economic development, create jobs, and improve the quality of life for individuals who were previously excluded from certain markets.

In summary, Market-creating game-changing innovation involves creating and serving markets that were previously excluded or untapped. Innovators in this space focus on meeting the needs of non-consumers, often through the introduction of affordable, accessible solutions. The impact of market-creating game-changing innovation extends beyond business success, influencing economic and social dynamics in profound ways.

2. Market-altering disruption

Market-altering disruption, another form of game-changing innovation, refers to the strategy of entering a market by providing products or services at significantly lower prices compared to existing offerings. This game-changing approach allows companies to attract a large customer base by appealing to cost-conscious consumers or those who were previously excluded due to high prices.

However, there are other examples of companies entering the market at the top end of the existing market. For example, Tesla entered the electric vehicle market with a high-end sports car.

Successful market-altering innovation often simplifies their offerings to reduce production and operational costs. This simplicity allows for streamlined processes, enabling companies to maintain profitability while offering lower prices. The strategy often involves targeting markets or customer segments that were underserved or overlooked by existing competitors. This can include price-sensitive consumers or those in emerging economies.

Market-altering game-changers often achieve cost efficiency through innovations in production processes, supply chain management, and distribution. Technological advancements may play a crucial role in facilitating these efficiencies.

Southwest disrupted the airline industry by offering low-cost, no-frills flights. The airline focused on operational efficiency, quick turnarounds, and a simplified business model to keep costs low. Dell disrupted the computer industry by selling directly to consumers and eliminating the need for a retail middleman. This direct-to-consumer model allowed Dell to offer competitively priced, customisable computers. Online platforms and e-commerce companies often disrupt traditional retail by operating with lower overhead costs and passing on the savings to consumers. This has been evident in various industries, including retail, travel, and entertainment.

Market-altering game-changers typically rely on a scale- and volume-based model. As they attract a larger customer base, they can achieve economies of scale, further driving down costs per unit. Penetrating the market with low prices is a key element of the strategy. This can lead to rapid customer acquisition and market share growth, often by capturing the market segment that previously purchased the higher-priced product or service from incumbents.

Market-altering game-changers must remain agile to sustain their competitive edge. As markets evolve, companies must continuously find new ways to reduce costs and improve efficiency. Market-altering game-changers often empower consumers by providing them with affordable choices. This can lead to increased competition, improved consumer welfare, and industry-wide innovation.

In summary, market-altering game-changing innovation is a strategic approach focusing on offering products or services at significantly lower prices in a new way. This game-changing strategy often requires innovation in operational efficiency and a deep understanding of consumer needs to succeed in the long term.

How to Become a Game Changer – Three Tests

Very few technologies or product ideas are inherently continuing or game-changing when they emerge from the inventors' minds. Instead, they go through a process of refinement into a strategic plan to win funding. Many ideas that get shaped into continuing innovation could just as easily be firmed up as game-changing innovations with much greater growth potential.

Innovators must answer three questions to determine whether an idea is a disruptive innovation or not.

1. Litmus tests for market-creating game-changers:

   a. Not served – Is there a large population of people who historically have not had the money, equipment, or skill to do things and, as a result, have gone without using it altogether?

   b. Inconvenient – Do they currently struggle to use the product or service?

If the technology can be developed so that a large population of less skilled or less affluent people can use the product or service, then there is potential for shaping the idea into a new-market disruption.

2. Litmus tests for market-altering game-changers:

   a. Over-performance – Are there customers at the lower end of the market who would be happy to purchase a product with fewer features and lower performance if they could get a lower price?

   b. Business model – Can we create a new business model that enables us to earn attractive profits at a discounted price to win the business of over-served customers?

3. Tests for both game-changing innovations:

   a. Industry-wide – Is the game-changing innovation applicable to all the incumbents in the industry?

If an idea fails any of the tests, it cannot be developed as a game-changing innovation. It may have the potential to become a continuing innovation.

Technology-based Game-changing Innovation

The emergence of technology-based innovations in the recent past is noteworthy. This has the potential to create both game-changing innovations and continuing innovations. Technology-based game-changing innovation refers to the introduction of new technologies that fundamentally transform industries, markets, and existing business models, often leading to the displacement of established products, services, or processes. These developments often fall into either market-creating or market-altering game-changers.

Game-changing innovations are typically driven by the introduction of new and often advanced technologies. These technologies have the potential to offer unique advantages, such as increased efficiency, cost-effectiveness, or novel features.

Technology-based game-changers often challenge existing business models by providing alternatives that are more efficient, accessible, or affordable, which leads to the inclusion of new customer segments previously not catered to. Incumbent businesses may struggle to adapt to these new models. Successful game-changing technologies tend to experience rapid adoption once they prove their value. This can result in a swift shift in market dynamics and consumer preferences.

The introduction of smartphones disrupted various industries, including telecommunications, photography, and navigation. It transformed how people communicate, access information, and consume media. Streaming platforms disrupted traditional cable and satellite television by offering on-demand, affordable, and customisable content delivery.

Technology-based disruptions often lead to the creation of new business ecosystems. New players and startups may emerge to capitalise on the opportunities presented by disruptive technology. Game-changing technologies can also influence consumer behaviour by providing new ways to access, interact with, or consume products and services. This shift may challenge traditional channels and methods.

Game-changing innovations are often iterative. As technologies evolve, subsequent iterations may bring additional game-changing or continuing innovation, leading to a continuous cycle of innovation.

While game-changing innovation is a process of evolution, many new and emerging technologies will disrupt existing business models. AI will play a critical part in development in the next century. It is already making inroads in many areas of our lives.

AI has the potential to make the game more competitive, making it a level playing field across all nations. Improving competitiveness and innovation are key ingredients to remain relevant in the future. Let's understand what it is and how it will impact our organisations!

3

Trajectory of Technology

Several laws and theories have been proposed over the years to explain the trajectory of technological growth. These laws often relate to the rate of technological advancement, the efficiency of technologies, and the adoption of new technologies.

There are three principles or laws that govern the trajectory of innovation or technological innovation. These principles enable us to predict the future in a more logical and scientific way. The below observations have now turned to generally accepted laws or principles that continue to drive technology and human advancement for the foreseeable future.

It is essential to understand the fundamentals before delving into a detailed discussion on AI since it depends on computing power.

1. Moore's Law: Computing power doubles about every two years.

The computing power doubles every two years due to improvements in computer chips. This means our computers can do tasks faster or handle more complex tasks previously not possible every two years. It has been known as Moore's Law since 1965, named after Gordon Moore.

Moore was an American businessman, engineer and co-founder of Intel Corporation, which pioneered the computer chips industry. His theory revolutionised technology in the 20st century.

We have witnessed this phenomenon for more than 50 years. The computer that navigated the "moon landing" mission in 1969 was about 0.045Mhz in speed compared to 4.19Mhz in the Nintendo Game Boy handheld game console. The first Apple iPhone launched in 2008 was significantly faster than those computers.

If this trend continues for the next 30 years, we will witness more powerful computers in our pockets. The increasing power will empower us to either do the same tasks faster or make more complex tasks more manageable. This will also enable more opportunities for AI as faster computing power will make complex algorithms possible.

2. Wright's Law: By doubling production, the cost falls by 10%–15%.

This law predicts the cost of producing gadgets such as computers and phones. It states that costs will fall by a constant percentage for every cumulative doubling of units produced. The rule of thumb is 10%–15%, thus making it a good predictor of the future cost of technologies. The theory has been known as Wright's Law since 1936.

Theodore Paul Wright was an American aeronautical engineer and educator. While studying aeroplane manufacturing, Wright determined that the labour requirement was reduced by 10–15% for every doubling of aeroplane production, which was the foundation for his seminal work later.

3. Amara's Law: Initial excitement followed by disappointment but eventual transformation.

Roy Amara, an American scientist and futurist, famously said, "We tend to overestimate the impact of technology in the short run but underestimate it in the long run." This is known as Amara's Law.

Amara's Law implies that between the early disappointment and the later underestimation, there must be a moment when we get it about right; pundits estimate it is 15 years down the line.

This is a very important law that must be read in conjunction with Moore's Law and Wright's Law. All three of them go hand in hand in understanding the future of technologies. It is important to note that they are observations and trends rather than immutable physical laws. Nevertheless, they offer a framework for anticipating how new technologies might develop and impact society and the economy.

Today, AI is one of the main beneficiaries of the evolution of technology over the last fifty years. It is clear that AI will continue to evolve due to the continued growth of computing power and reduction in cost that make technology accessible to most if not all. The impending Quantum Computing revolution is expected to future revolutionise this trend in the future.

4

Artificial Intelligence

The launch of ChatGPT in November 2022 marked the beginning of the emergence of generative AI tools for consumers. It quickly became the fastest-growing consumer app when it reached 100 million users within a couple of months of its launch. ChatGPT was trained on a much larger dataset than its predecessors, with far more parameters. ChatGPT is believed to use trillions of data parameters to train its algorithms. It is a technology-based game-changing innovation that has the potential to create new disruptive business models and provide continuing innovation to other industries.

Definition

Artificial Intelligence (AI) is the simulation of human intelligence in machines programmed to think, learn, and perform tasks commonly associated with human intelligence. In other words, it is machines and algorithms that think like humans, act like humans, and behave like humans.

These tasks include problem-solving, understanding language, recognising patterns, speech recognition, and visual perception. The goal of AI is to develop systems that can perform tasks autonomously, improving their performance over time through experience and learning.

According to the latest revision of the EU *AI Act*, AI is defined as "a machine-based system that is designed to operate with varying levels of autonomy, and that can, for explicit or implicit objectives, generate outputs such as predictions, recommendations, or decisions that influence physical or virtual environments."

Examples of AI Applications

AI has become increasingly integrated into the finance profession, offering advanced tools and solutions in an attempt to streamline processes, enhance decision-making, and improve overall efficiency. Below are various AI applications within the finance profession.

1. Algorithmic trading

AI-driven algorithms analyse market data, identify patterns, and execute trades at high speeds, thereby optimising trading strategies. This minimises emotional decision-making and enables swift responses to market changes.

2. Portfolio management

AI-powered systems assist in portfolio construction, optimisation, and rebalancing based on risk tolerance, market conditions, and financial goals. It enhances portfolio performance, tailers investment strategies to individual preferences, and automates decision-making.

3. Credit scoring and risk assessment

AI algorithms analyse extensive data to assess creditworthiness, predict default risks, and inform lending decisions. This application improves accuracy in credit scoring, enhances risk prediction models, and aids in better loan underwriting.

4. Fraud detection and prevention

AI systems analyse transaction patterns, detect anomalies, and identify potentially fraudulent activities in real-time. It enhances security, reduces financial losses, and provides warnings for suspicious activities.

5. Customer service chatbots

AI-powered chatbots handle customer inquiries, provide account information, and assist with routine tasks, with the intention of improving customer service. This reduces response times and handles a large volume of inquiries simultaneously. Many large organisations have implemented AI-powered chatbots in accounts payable, procurement and general accounting to automate routine and simple tasks.

6. Financial advisory

AI-driven advisory platforms provide personalised financial advice, investment recommendations, and retirement planning, empowering clients to make informed decisions.

7. Regulatory compliance and reporting

AI assists in monitoring and ensuring compliance with financial regulations by automating compliance processes and analysing large datasets, thus reducing the risk of regulatory violations, enhancing transparency, and streamlining reporting requirements.

8. Invoice and receipt processing

AI automates the extraction of data from invoices and receipts, streamlining the accounts payable process. Automation increases processing speed, reduces errors, and allows finance professionals to focus on more strategic tasks.

9. Predictive analytics for financial markets

AI models analyse historical data and market trends to make predictions about future market movements and investment opportunities. This aids in strategic decision-making, improves investment strategies, and helps risk management.

10. Automated compliance audits

AI tools automate the auditing process by analysing financial transactions and ensuring adherence to regulatory standards. The automation improves accuracy, reduces the time required for audits, and enhances overall compliance.

•

While the above are examples of legitimate AI applications, there remain many examples of misuse. For example, the use of bots in election campaigns across the world has threatened to destabilise the democratic process. These cases demand a robust and comprehensive framework for AI ethics before it goes out of control.

As the finance industry continues to adopt AI, professionals can leverage these applications to enhance productivity, improve decision-making, and stay competitive in an ever-evolving landscape. Additionally, AI-driven tools enable finance professionals to focus on more strategic tasks, contributing to the overall growth and innovation in finance.

Types of AI

AI can be classified into two main types: Specific AI and General AI.

Specific AI, also known as Narrow AI and Weak AI, is designed to perform a specific task or a set of closely related tasks. It operates within a well-defined context and performs a particular function. Examples include virtual personal assistants, image recognition software, and recommendation algorithms.

General AI, while still a theoretical concept, refers to an AI system that can understand, learn, and apply knowledge across various tasks, akin to human abilities. True General AI does not exist today, and AI systems are primarily developed for specific applications. However, there is growing concern that AI systems soon will have human-like capabilities that might undermine the very existence of humanity.

AI Techniques

AI systems employ various techniques and approaches. These include but are not limited to the below.

1. Machine Learning

Machine Learning (ML) is a powerful subset of AI that can learn patterns and make predictions without explicit programming. At its core, ML relies on algorithms that iteratively learn from data, enabling computers to recognise patterns, uncover hidden insights, and make decisions or predictions based on that acquired knowledge.

The process involves feeding the algorithm a training dataset containing input features and corresponding output labels. The model learns to map the input data to the correct output by adjusting its parameters through iterative training. Supervised learning involves using labelled data, where the model is trained on known input-output pairs, while unsupervised learning involves extracting patterns and relationships from unlabelled data. Reinforcement learning, another ML paradigm, focuses on training agents to make sequential decisions by receiving feedback in the form of rewards or penalties, allowing them to learn optimal strategies through interaction with an environment.

ML finds application across diverse domains, revolutionising industries and enhancing decision-making processes. From image and speech recognition to predictive analytics, ML is at the forefront of technological innovation. In healthcare, it aids in diagnosing diseases based on medical images and patient data. In finance, ML algorithms detect fraudulent transactions and make predictions on stock market trends. In marketing, recommendation systems leverage ML to suggest products tailored to individual preferences.

The versatility of ML continues to drive advancements, making it an indispensable tool for addressing complex challenges and unlocking new possibilities in the realms of technology, science, and business. As the field continues to evolve, the integration of ML into various facets of our lives is poised to reshape how we interact with and harness the power of technology.

2. Deep Learning

Deep Learning (DL) represents a sophisticated and powerful subset of ML that revolves around artificial neural networks with multiple layers, often referred to as deep neural networks.

Unlike traditional ML algorithms where data is labelled and structured for ML using feature engineering and manual extraction, DL models automatically learn hierarchical representations of data through these intricate neural networks. These deep architectures enable the systems to recognise intricate patterns and features in complex datasets, making them particularly effective in tasks such as image and speech recognition, colloquial language recognition or natural language processing (NLP), and even playing strategic games like Go. The depth of these networks allows them to capture and understand intricate nuances within the data, making DL a transformative force in the field of AI.

One of the hallmark achievements of DL is its ability to handle large volumes of unstructured data. Convolutional Neural Networks (CNNs), a specific class of deep neural networks, have revolutionised image processing by autonomously learning hierarchical features, enabling accurate image recognition and object detection.

Recurrent Neural Networks (RNNs) are adept at handling sequential data, making them indispensable in natural language processing and speech recognition tasks. The success of DL is evident in its widespread applications across industries, including healthcare for medical image analysis, autonomous vehicles for object detection, and language translation services. As computational power and data availability continue to grow, DL is poised to lead the way in unravelling the complexities of real-world data and driving further advancements in AI.

3. Generative AI

Generative AI (Gen AI), at the forefront of AI innovation, refers to a class of systems designed to produce new, original content autonomously. Generative AI is a subset of DL.

Among the various techniques within generative AI, Generative Adversarial Networks (GANs) stand out. GANs employ a dual-network structure where a generator and a discriminator are pitted against each other in a constant feedback loop. The generator generates synthetic data while the discriminator evaluates its authenticity. Through this adversarial training, GANs excel in tasks like image synthesis, enabling the creation of remarkably realistic images that can be indistinguishable from those captured in the real world. This technology has found applications in art, design, and entertainment, pushing the boundaries of computer-generated content.

Another facet of generative AI lies in text generation models, where Recurrent Neural Networks (RNNs) and transformer-based architectures like Generative Pre-trained Transformers (GPT) have made significant strides. These models learn the intricacies of language patterns from vast datasets and can generate coherent and contextually relevant textual content. Applications range from natural language interfaces in chatbots to creative writing assistance and even the generation of news articles.

Generative AI's ability to create content across various modalities, from images to text, showcases its versatility and potential impact on industries seeking innovative ways to generate new, meaningful content.

However, the ethical considerations surrounding its applications, such as deepfake generation, highlight the importance of responsible development and deployment of generative AI technologies. For example, deepfakes have been used in audio, photography, and videos often without consent to create content passed off as human-generated. Artists generating artwork using AI and writers leveraging GPT tools to write content are topics debated frequently in the public domain. Sora, OpenAI's latest generative AI tool that generates video from text, promises to understand and simulate the physical world in motion, with the goal of training models that help people solve problems that require real-world interaction. Sora can generate videos up to a minute long while maintaining visual quality and adherence to the user's prompt is a game-changing innovation in creative design.

4. Discriminative AI

Discriminative AI models are a category of ML algorithms focused on understanding and directly modelling the boundary between different classes or categories within a dataset.

The primary objective of discriminative models is to learn the conditional probability distribution of the output given the input. In other words, they aim to discern the features that distinctly separate different classes or categories in the data. Commonly used discriminative models include logistic regression, support vector machines, and neural networks with a SoftMax output layer. These models are particularly well-suited for classification tasks where the goal is to accurately assign input data to predefined categories. Discriminative AI has found extensive applications in various domains, ranging from image and speech recognition to sentiment analysis and medical diagnosis, where the emphasis is on distinguishing between different outcomes or classes.

While generative AI generates new content such as images and text, discriminative AI distinguishes them based on input attributes. For example, generative AI can generate a picture of a dog; discriminative AI will identify a dog based on input data.

5. Large Language Models

A subset of DL, Large Language Models (LLMs) represent a groundbreaking advancement in the field of natural language processing. These models are characterised by their vast neural network architectures, often containing billions or even trillions of parameters. One of the notable examples is OpenAI's GPT-4, which is one of the largest language models to date with about 170 trillion parameters. The sheer size of these models allows them to capture intricate nuances of language, enabling them to understand context, generate coherent text, and perform a myriad of language-related tasks.

LLMs are typically pre-trained on massive datasets containing diverse linguistic patterns and information, and they can subsequently be fine-tuned for specific applications, making them versatile tools for various industries and applications.

LLMs have demonstrated exceptional performance across a wide range of natural language tasks. They excel in language generation, translation, summarisation, question and answering, and even creative writing. The ability of LLMs to comprehend context and generate contextually relevant responses has elevated their role in conversational AI, virtual assistants, content creation, and more.

However, their massive scale also raises concerns related to computational resources, energy consumption, and ethical considerations. Striking a balance between the remarkable capabilities of large language models and the responsible use of resources remains a topic of ongoing research and discussion in the AI community.

Figure 1 below depicts how all of the AI techniques fit together.

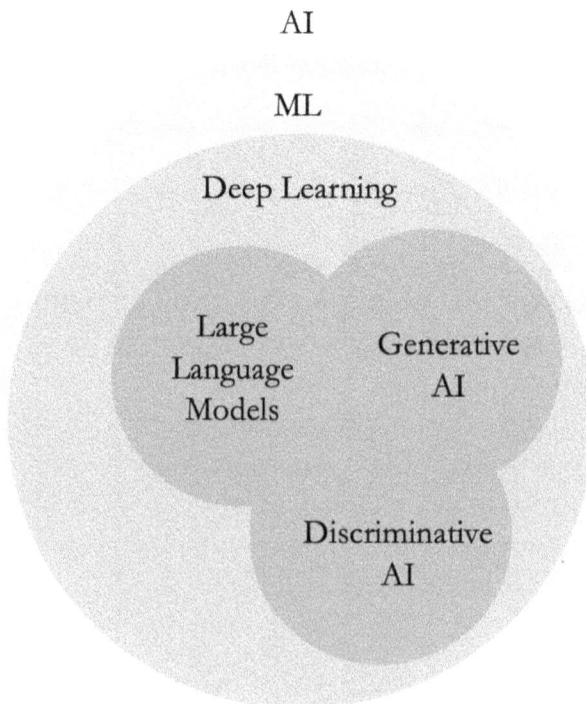

AI

ML

Deep Learning

Large Language Models

Generative AI

Discriminative AI

Figure 1

AI Learning

AI learning focuses on the development of algorithms and models that enable computers to learn and make decisions based on data without being explicitly programmed. AI learning involves training models on large datasets, allowing them to identify patterns, recognise trends, and make predictions or decisions in new situations.

There are several key types of AI learning methods, each serving different purposes.

1. Supervised learning

Supervised learning is a foundational and widely employed paradigm within the field of ML, where algorithms are trained on labelled datasets to make predictions or classifications.

In supervised learning, the model is provided with input data along with corresponding output labels, forming a training set. The algorithm aims to learn the underlying patterns and relationships between the input features and the desired outputs. Through this learning process, the model generalises its understanding and becomes capable of making accurate predictions on new, previously unseen data. Common supervised learning applications include image and speech recognition, natural language processing, and predictive modelling in various domains.

The workflow of supervised learning involves the following key steps: data collection, where a labelled dataset is curated; data preprocessing, in which the data is cleaned and transformed to make it suitable for training; model training, where the algorithm adjusts its parameters based on the labelled data; and model evaluation, which assesses the model's performance on a separate dataset not used during training. The success of supervised learning hinges on the availability of high-quality labelled data, as it enables the model to learn and generalise from the provided examples. Despite its effectiveness, supervised learning is limited by the need for extensive labelled datasets and may struggle with tasks where labelled data is scarce or expensive since it needs to be manually labelled.

Supervised learning finds application in many real-world scenarios, demonstrating its versatility and effectiveness across diverse domains. In healthcare, supervised learning algorithms are employed for disease diagnosis based on medical imaging data, such as detecting cancer and tumours in radiological scans. In finance, these algorithms excel at credit scoring, assessing the creditworthiness of individuals by learning patterns from historical data.

Natural language processing tasks, including sentiment analysis and language translation, leverage supervised learning to understand and generate human-like text. Supervised learning powers facial recognition systems in computer vision, enabling accurate identification in security and authentication applications. Additionally, recommendation systems in e-commerce and streaming platforms use supervised learning to predict user preferences and provide personalised content suggestions. The widespread adoption of supervised learning underscores its effectiveness in solving a diverse array of problems where labelled data is readily available.

2. Unsupervised learning

Unsupervised learning is a transformative paradigm within ML, where algorithms delve into datasets without explicit guidance in the form of labelled data. The primary objective of unsupervised learning is to uncover inherent patterns, structures, or relationships within the data, paving the way for insights and discoveries. Clustering is a prominent application of unsupervised learning, where algorithms group similar data points based on shared characteristics. This is exemplified in business customer segmentation, where unsupervised learning helps identify distinct customer groups with similar purchasing behaviours without predefined labels.

Another significant application of unsupervised learning is dimensionality reduction, a technique that simplifies the complexity of data by identifying and preserving its essential features. Principal Component Analysis (PCA) is a classic example of compressing high-dimensional data into a lower-dimensional representation while retaining its salient information.

Unsupervised learning's ability to extract meaningful patterns from unlabelled data is particularly valuable in scenarios where obtaining labelled data is impractical or costly.

Anomaly detection, a critical application in cybersecurity, leverages unsupervised learning to identify deviations from the norm within network traffic or user behaviour, enabling swift detection of potential security threats. Additionally, unsupervised learning plays a pivotal role in exploratory data analysis, helping researchers and analysts uncover hidden structures in large datasets without preconceived notions. Whether unravelling patterns in financial transactions to detect fraudulent activities or segregating components in multimedia data, unsupervised learning stands as a versatile tool in AI.

3. Semi-supervised learning

Semi-supervised learning stands as an intermediate paradigm between fully labelled and unlabelled learning. In scenarios where obtaining extensive labelled datasets is challenging or expensive, semi-supervised learning proves to be a pragmatic solution.

The model initially learns from the limited labelled examples, capturing patterns and relationships in the explicitly provided data. Its training is then complemented by a wealth of unlabelled data, allowing the algorithm to generalise and adapt to a broader range of situations.

One notable advantage of semi-supervised learning is its applicability to real-world scenarios where obtaining fully labelled datasets may be impractical. For example, in natural language processing, semi-supervised learning techniques can be employed for sentiment analysis, where a smaller set of labelled reviews can be used to train a model that can then analyse and categorise a vast amount of unlabelled text data. Another example is medical imaging where a small sample is enough to initially train the model.

This hybrid learning approach addresses challenges associated with data labelling and enhances the model's adaptability and generalisation capabilities, making it a valuable tool in contexts where comprehensively labelled datasets are scarce.

4. Reinforcement learning

Reinforcement learning is a dynamic paradigm within ML, characterised by agents that learn optimal decision-making strategies through interactions with an environment.

Unlike supervised learning, where models are trained on labelled datasets, reinforcement learning involves an agent taking actions within an environment and receiving feedback as rewards or penalties. The objective is for the agent to learn a policy – a set of rules or strategies – that maximises the cumulative reward over time. This learning process is often modelled as a Markov Decision Process (MDP), where the agent perceives the state of the environment, takes an action, receives a reward, and transitions to a new state. Notable applications of reinforcement learning include robotics, where agents learn to perform complex tasks and game playing, exemplified by algorithms mastering strategic games like Go and chess.

One of the key challenges in reinforcement learning is striking a balance between exploration and exploitation. The agent must explore different actions to discover the most rewarding ones while also exploiting its current knowledge to make decisions that yield immediate rewards.

Techniques such as Q-learning and policy gradient methods are employed to train agents to navigate various environments. Reinforcement learning's adaptability and capacity to handle sequential decision-making scenarios make it well-suited for applications in autonomous systems, recommendation systems, and even in optimising resource allocation in areas like finance and logistics. As research in reinforcement learning progresses, it continues to contribute to advancements in AI, robotics, and decision-making systems.

AI Model Architecture

AI model architecture refers to the structure and organisation of the underlying framework that enables an AI model to process information, make decisions, and generate outputs. The architecture defines the configuration and connectivity of the model's components, including layers, nodes, and parameters.

Different AI tasks, such as image recognition, natural language processing, and reinforcement learning, often require specific architectures tailored to the nature of the data and the complexity of the task. Below are brief overviews of some common AI model architectures.

1. Neural networks

Neural networks form the backbone of modern AI and ML systems, drawing inspiration from the structure of the human brain. Composed of interconnected nodes organised into layers, neural networks are designed to process information, identify patterns, and make predictions.

The basic unit, or neuron, mimics a biological neuron by receiving inputs, applying a weighted sum, and passing the result through an activation function to produce an output. Stacking these neurons into layers allows neural networks to model complex relationships within data. An input layer receives the initial data, hidden layers process this information through weighted connections, and an output layer produces the final prediction or classification.

Deep neural networks, often referred to as DL models, extend the concept of neural networks by incorporating multiple hidden layers. This depth enables them to learn intricate hierarchical representations of data, making them particularly effective in tasks such as image recognition, natural language processing, and speech recognition. Convolutional Neural Networks (CNNs) are specialised architectures within DL, tailored for processing grid-like data, such as images. Recurrent Neural Networks (RNNs) are designed for sequential data, where the network maintains memory of past inputs, making them suitable for tasks like language modelling and time series analysis.

The training process of neural networks involves adjusting the weights of connections through a mechanism called backpropagation. During this process, the network learns from labelled examples by minimising the difference between its predictions and the actual outcomes. Neural networks have demonstrated remarkable success in various domains, from image and speech recognition to natural language understanding, contributing significantly to the advancements in AI.

2. Transformer networks

Transformer networks have emerged as a revolutionary architecture in the field of natural language processing and ML. Introduced in the paper "Attention is All You Need", transformers have proven highly effective in capturing long-range dependencies and contextual information within sequences of data.

Unlike Recurrent Neural Networks (RNNs), transformers employ self-attention mechanisms, allowing each element in the input sequence to focus on different parts of the sequence during processing. This parallelised attention mechanism facilitates more efficient learning of relationships and dependencies, making transformers particularly powerful for tasks like language translation, text summarisation, and question and answer.

As transformer networks can process input data in parallel, they are highly scalable and well-suited for tasks involving large datasets.

This parallelisation enables transformers to capture global dependencies and contextual information without being hindered by the sequential nature of traditional recurrent networks. The architecture has been successfully applied in models such as Google BERT (Bidirectional Encoder Representations from Transformers) for contextualised language understanding and ChatGPT for language generation.

Transformers have significantly influenced the landscape of DL, demonstrating state-of-the-art performance in natural language processing benchmarks. Their versatility extends beyond language tasks, as they have been adapted for computer vision tasks, such as image recognition and object detection. The attention mechanism in transformers has become a fundamental building block in modern neural network architectures, showcasing the impact and adaptability of this innovative model.

•

AI stands at the forefront of technological innovation, reshaping the way we interact with information, automate tasks, and solve complex problems.

The evolution of AI has seen the rise of the versatility of the 21st-century enterprise. These advancements have empowered AI to excel in image recognition, understanding language, and strategic decision-making tasks. As AI continues to permeate various aspects of our lives and industries, ethical considerations and responsible development are crucial to its positive impact on society.

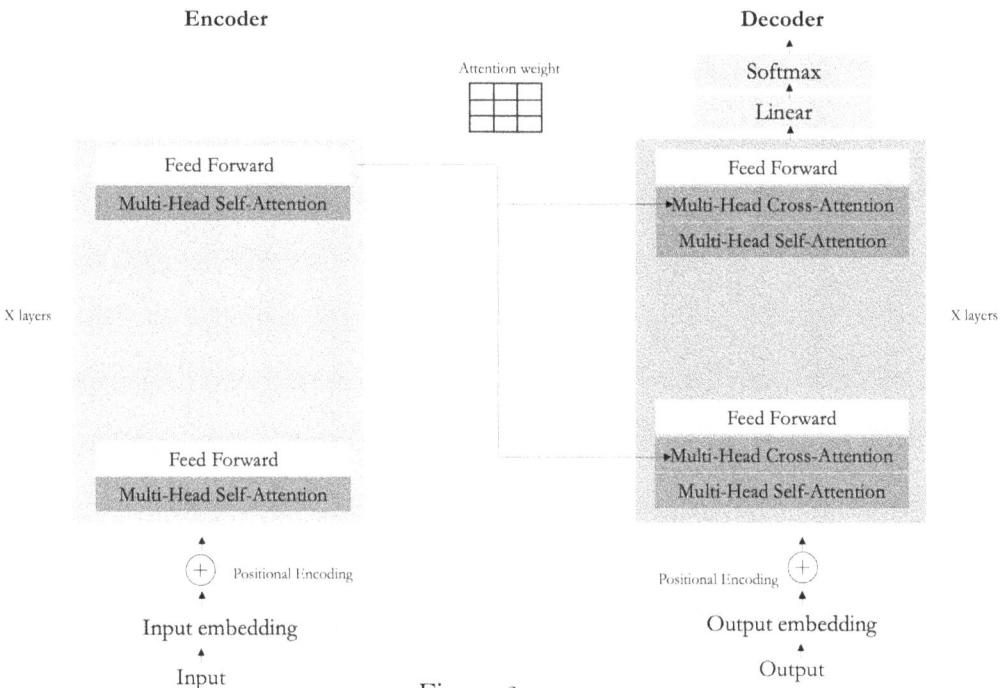

Figure 2

The journey of AI is dynamic, marked by ongoing research, breakthroughs, and ethical discourse. The promise of AI lies not just in its computational prowess but in its potential to augment human capabilities, drive innovation, and contribute to a future where intelligent systems collaborate with humanity for the betterment of our global community.

5

Components of AI Ecosystem

Understanding the high-level concepts of AI is becoming increasingly important for professionals in various fields, including accountancy. While finance professionals may not need to delve into the technical intricacies of AI components, having a solid high-level understanding allows them to navigate the evolving landscape of technology and make informed decisions about its integration into financial processes.

Figure 3 below depicts the components of an AI ecosystem. It is important to note that the figure does not represent a technical architectural diagram.

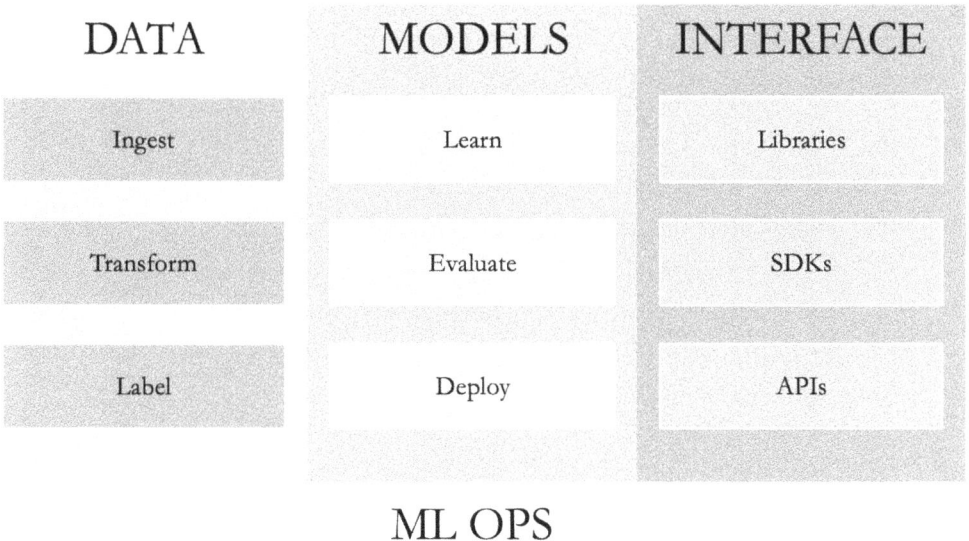

DATA	MODELS	INTERFACE
Ingest	Learn	Libraries
Transform	Evaluate	SDKs
Label	Deploy	APIs

ML OPS

Figure 3

Below are some key impacts of AI on organisations. These are potentially beneficial if executed appropriately.

1. Informed decision-making

Finance professionals often play a crucial role in decision-making within organisations. A high-level understanding of AI enables them to assess the potential impact of AI technologies on financial workflows, facilitating more informed decisions regarding finance and adopting AI tools.

2. Efficiency and automation

AI has the potential to automate routine and repetitive tasks, thereby enhancing efficiency in financial operations. Accountants who grasp the high-level concepts of AI can identify opportunities for automation, streamline processes, and allocate resources more effectively.

3. Data-driven insights

AI, particularly in data analytics, can provide valuable financial analysis and reporting insights. Finance professionals with a high-level understanding can leverage these tools to extract meaningful information from large datasets, enabling more accurate financial forecasting and decision support.

4. Compliance and risk management

As AI is integrated into various business functions, understanding the high-level implications of AI on compliance and risk management is crucial. Finance professionals can contribute to ensuring that AI applications adhere to regulatory standards and mitigate potential risks associated with data privacy and security.

5. Collaboration with tech professionals

While accountants may not be responsible for developing AI systems, a basic understanding allows for more effective collaboration with technical professionals. Clear communication and comprehension of AI concepts facilitate collaboration between accountants, data scientists, and AI specialists when implementing AI solutions.

6. Continuous professional development

The field of AI is dynamic, with advancements occurring rapidly. Finance professionals with a high-level understanding can engage in continuous learning, staying abreast of emerging trends and technologies. This adaptability is crucial in an age where technology is transforming various aspects of business.

•

While not necessary to become AI experts, this knowledge and basic understanding equips finance professionals such as accountants to navigate the evolving landscape of technology-driven financial processes effectively.

6

Data

Data refers to raw and unprocessed information or facts that are collected and stored.

Its effective utilisation is essential across various fields and industries. The interpretation and transformation of data into meaningful information contribute to informed decision-making, problem-solving, and the development of insights and knowledge.

In computing and information technology, data is typically organised and structured to enable efficient storage, retrieval, and analysis.

Types of Data

Data comes in various forms, and examples span various formats and sources. Below are examples of different types of data.

1. Numerical data

Numerical data consists of quantitative information expressed in numerical form, typically consisting of numbers. This data category is fundamental to various fields, including science, economics, and engineering, as it enables precise calculations, statistical analysis, and mathematical modelling.

Numerical data can be further categorised into discrete, where the values are distinct and separate (e.g., the number of students in a class), and continuous, where the values form a continuous range (e.g., temperature readings). Whether representing measurements, quantities, or mathematical values, numerical data provides a quantitative foundation for analytical processes, aiding in decision-making, trend identification, and the formulation of mathematical models in diverse domains.

2. Text data

Text data encompasses alphanumeric characters arranged to form words, sentences, or paragraphs, constituting a fundamental and versatile type of information. Often used in communication, documentation, and literature, text data is ubiquitous in the digital age.

This category includes everything from social media posts and emails to articles, books, and more. Textual information is expressive and a rich source for natural language processing and sentiment analysis, allowing for the extraction of insights from written communication. Analysing text data involves techniques such as text mining, which seeks patterns and trends within a vast textual content. Whether used for understanding customer feedback, sentiment analysis on social media, or automating information extraction, text data plays a crucial role in shaping our understanding of language and communication in the digital landscape.

3. Image data

Image data represents visual information through pixel values, forming a fundamental component in computer vision and image processing.

Comprising intricate patterns of colour and intensity, image data is used to capture and convey visual scenes, making it crucial in various applications. From medical imaging and satellite photography to facial recognition and augmented reality, image data enables machines to interpret and understand visual content.

Each pixel's colour and position contribute to the overall composition of an image, allowing for the extraction of features and patterns. Analysing image data often involves advanced techniques such as

Convolutional Neural Networks (CNNs), which excel at recognising spatial hierarchies within visual data. As a cornerstone in computer vision, image data is pivotal for tasks ranging from object detection to image classification, shaping the way machines perceive and interpret the visual world.

4. Audio data

Audio data comprises waveforms that capture variations in air pressure over time, translating into sound. This form of data is central to the field of audio processing and is encountered in diverse applications, from music streaming services to speech recognition systems.

Waveforms are typically represented digitally, allowing for manipulation, analysis, and playback by electronic devices. The richness of audio data lies in its ability to convey not only musical compositions but also spoken language, ambient sounds, and environmental noises. Analysing and processing audio data involve techniques such as Fourier transforms for frequency analysis and ML models for speech-to-text conversion or music recommendation. As technology advances, the significance of audio data continues to grow, influencing innovations in voice-controlled devices, personalised audio experiences, and advancements in the broader field of AI.

5. Video data

Video data is a dynamic and sequential representation of visual information, composed of a series of individual frames displayed rapidly to create the perception of motion. This form of data is integral to numerous applications, ranging from entertainment and surveillance to education and scientific research. Each frame within a video holds valuable visual content, and the sequential nature of video data allows for the capture of temporal patterns and dynamic events.

Analysing video data involves sophisticated techniques in computer vision, ML, and image processing, enabling tasks such as object recognition, activity tracking, and scene understanding. As technology advances, video data continues to play a crucial role in shaping applications like video streaming, virtual reality, and automated video content analysis, contributing to a richer and more immersive understanding of the visual world.

6. *Geospatial data*

Geospatial data refers to information that relates to specific geographic locations on the Earth's surface. It encompasses a diverse range of data types, including geographic coordinates, satellite imagery, maps, and spatial relationships. This type of data plays a crucial role in various fields such as mapping, navigation systems, urban planning, environmental monitoring, and disaster response.

Geospatial data allows for the analysis of spatial patterns, trends, and interactions, providing valuable insights for decision-making processes. Advanced technologies, such as geographic information systems (GIS), facilitate the collection, storage, and visualisation of geospatial data, enabling professionals and researchers to better understand the complexities of the physical world and make informed choices based on location-specific information.

Categories of Data

Data can be categorised into several types based on different criteria. Here are some common categories of data.

1. *Structured data*

Structured data refers to information that is organised and formatted in a way that fits neatly into predefined data models. This type of data is characterised by its clear and fixed structure, often arranged in tables with rows and columns.

Structured data is commonly found in relational databases and is easily queried using standard database management systems. Examples of structured data include numeric data such as integers or decimals, dates, and categorical data like product categories.

Given the relationships between data elements are defined, such structured data can be compatible across various software systems for storage, retrieval, processing, and analysis.

In the example below (Figure 4), each row represents an employee, and each column represents a specific attribute such as Employee ID, First Name, Last Name, Department, Salary, and Hire Date. The structured format allows for easy organisation and retrieval of information, making it suitable for tasks like querying for specific employee details, sorting based on salary, or filtering by department.

EmployeeID	First Name	Last Name	Department	Salary	Hire Data
001	John	Smith	Sales	$60,000	15/01/2022
002	Mary	Robinson	Finance	$55,000	22/08/2021
003	Rob	France	Marketing	$70,000	10/03/2023
004	Emily	Paris	HR	$65,000	6/12/2020
005	Michael	Lamb	IT	$75,000	26/08/2019

Figure 4

2. Unstructured data

Unstructured data refers to information that is not in a predefined and organised format, often existing in a free-form or narrative structure. Unlike structured data which is organised into tables or databases, unstructured data comes in diverse and dynamic formats, including text documents, emails, images, audio recordings, and video files. This data type poses challenges for traditional data processing methods due to its variable and flexible formats.

Extracting meaningful insights from unstructured data typically involves advanced technologies such as natural language processing (NLP), computer vision, and ML algorithms. Despite its inherent complexity, unstructured data holds valuable information, providing a rich source for understanding sentiments, extracting patterns, and gaining qualitative insights from sources like customer feedback, social media interactions, and other forms of human communication. As organisations increasingly encounter vast amounts of unstructured data, harnessing its potential requires innovative approaches to unlock valuable knowledge hidden within the diversity of formats and sources.

The data sample below (Figure 5) presents the information in a narrative-like form without a predefined and organised structure. It includes details of customer feedback, action items, and plans for the next meeting. To most people, there is recognisable structure. It reads like minutes, but to a machine, there is no rigid structure. Analysing such data often requires natural language processing techniques to extract valuable insights.

Attendees: - Jane Smith (Customer)
- Mark Johnson (Sales Representative)
- Sarah James (Customer Support)

Feedback Highlights:
- Jane expressed satisfaction with the product's ease of use but suggested adding more tutorial videos.
- Mark reported positive feedback from several clients regarding the new features introduced last month.
- Sarah shared customer concerns about a recent service outage; proposed improvements were discussed.

Action Items:
1. Marketing to create additional tutorial videos for the product.
2. 2. The sales team to gather more detailed feedback from clients on recent features.
3. 3. The IT department to address and improve system stability based on customer support feedback. Next Meeting: December 15, 2023

Figure 5

3. Semi-structured data

Semi-structured data represents a middle ground between fully structured and unstructured data, embodying elements of both. Unlike structured data with a rigid, tabular format, semi-structured data retains some flexibility in its organisation. It often adheres to a predefined schema but allows for variations within that structure.

Data Sources

Data sources refer to the origins or locations from which data is collected or generated. In the modern digital landscape, diverse data sources contribute to the vast pool of information available for analysis and decision-making. Below are some common data sources.

1. Data from operational system

Data from operational systems encompasses the information generated and stored in the day-to-day functioning of an organisation's core processes. These systems, often referred to as operational or transactional systems, include databases, enterprise resource planning (ERP) software, customer relationship management (CRM) systems, and other applications crucial to the organisation's operations.

The data within these systems captures real-time transactions, interactions, and activities, reflecting the heartbeat of the business. Examples include sales transactions, inventory movements, customer orders, and employee records. Analysing data from operational systems provides insights into the efficiency, performance, and overall health of the organisation's processes. It enables timely decision-making, process optimisation, and the identification of areas for improvement. As organisations increasingly recognise the value of data-driven decision-making, leveraging insights from operational systems becomes integral to achieving strategic goals and maintaining a competitive edge in today's dynamic business environment.

An example of data in operational systems can be found in a retail business's point-of-sale (POS) system. The operational system is responsible for handling daily sales transactions and managing inventory.

2. Social media feeds

Social media data, which is mostly unstructured data, encompasses the vast and dynamic information generated across various social platforms by users engaging in online interactions. This data includes text, images, videos, likes, shares, and more, reflecting diverse aspects of user behaviour and preferences. Social media platforms such as Facebook, Twitter (now X), Instagram, TikTok, and LinkedIn serve as rich sources of real-time, user-generated content.

Analysing social media data provides valuable insights into trends, sentiments, and public opinion. Businesses leverage this data to understand customer feedback, tailor marketing strategies, and enhance brand engagement. Governments sometimes use this data to carry out surveillance on their citizens. Researchers use social media data for studying societal trends, sentiment analysis, and even predicting events.

However, managing and extracting meaningful insights from the sheer volume and unstructured nature of social media data often requires advanced techniques, including natural language processing and ML algorithms.

3. Public data

Public data or open-source data refers to information that is openly available to the public and can be accessed without restrictions. This data is typically provided by government agencies, research institutions, or other entities with a commitment to transparency and accessibility.

Public data encompasses a wide range of information, including demographic statistics, economic indicators, environmental measurements, and government reports. Initiatives such as open data platforms and government transparency efforts aim to make relevant and valuable datasets accessible for analysis and use by individuals, businesses, researchers, and policymakers. Public data plays a crucial role in fostering accountability, promoting research, and empowering citizens with information about various aspects of society. As the demand for data-driven decision-making continues to grow, the availability of public data serves as a cornerstone for informed insights and evidence-based decision-making across diverse fields.

4. Web streaming data

Web streaming data refers to the continuous flow of real-time information transmitted over the internet and is instantly accessible to users. Unlike static data, which remains unchanged until updated, web streaming data is dynamic and constantly evolving. This type of data is commonly associated with web clicks, live events, social media feeds, financial market updates, and other sources that require immediate dissemination of information.

Web streaming technologies allow users to consume content in real-time, enabling them to stay updated on events as they unfold. Analysing web streaming data presents unique challenges due to its rapid pace and volume, requiring specialised tools and techniques for real-time processing and decision-making. Businesses often leverage web streaming data to monitor online conversations, track trends, and respond promptly to emerging opportunities or issues in the digital landscape. The dynamic nature of web streaming data underscores its significance in providing timely insights and enhancing the responsiveness of applications and services in the interconnected world of the internet.

5. Surveys and questionnaires

Surveys and questionnaire data are structured sets of information collected through systematic inquiries of individuals or groups. Surveys are used to gather specific data points, opinions, or feedback on a particular subject. These instruments can be deployed through various mediums, including paper, online platforms, or in-person interviews. The data collected typically includes responses to specific questions, demographic information, and, in some cases, open-ended comments.

The structured nature of survey data allows for statistical analysis, making it valuable for interpretation and drawing insights from large respondent groups. Researchers, businesses, and policymakers use survey and questionnaire data to understand public opinion, consumer preferences, employee satisfaction, and a myriad of other topics. Effectively analysing this data involves not only quantitative methods but also qualitative approaches to gain deeper insights from open-ended responses, providing a comprehensive understanding of the surveyed subject.

6. Data from sensors and IoT devices

Data from sensors and Internet of Things (IoT) devices represent a crucial aspect of the modern digital landscape, providing a continuous stream of real-time information from the physical world. Sensors, ranging from temperature and humidity sensors to motion detectors, collect data on environmental conditions and physical phenomena. IoT devices, which are connected to the internet, encompass a diverse array of interconnected gadgets, from smart home devices to industrial sensors.

The data collected and generated by these devices offer insights into operational efficiency, environmental conditions, and user behaviour. In industrial settings, IoT devices may monitor machinery performance, optimising production processes and minimising downtime. In smart cities, sensors can gather data on traffic flow, air quality, and energy consumption for improved urban planning. Analysing data from sensors and IoT devices requires expertise in data science and often involves techniques such as ML to derive meaningful patterns and insights, making it a key component in the broader landscape of big data analytics.

7. Logs and server data

Logs and server data are integral components of IT infrastructure, providing detailed records of activities and events within computer systems and networks. Server logs, generated by servers and network devices, capture information such as user access, system errors, security events, and application performance metrics.

Since server data includes information on resource access and usage, network traffic, and user interactions, it offers valuable insights for security issues, capacity planning, and infrastructure optimisation. As organisations increasingly rely on digital technologies, leveraging logs and server data becomes essential for maintaining the reliability, security, and efficiency of their IT systems. Advanced log analysis tools and techniques, including ML, are often employed to derive meaningful insights from the vast amounts of data generated by servers and network devices.

8. Manual data

Manual data refers to information that is recorded, processed, or managed by humans without the use of automated systems or technology. In a manual data environment, individuals physically input, organise, and maintain data using traditional tools such as paper, notebooks, or basic spreadsheets.

While manual data handling may be suitable for small-scale or less complex operations, it often poses challenges related to efficiency, accuracy, and scalability. Human errors, transcription mistakes, and limited processing speed are inherent risks associated with manual data processes. As organisations evolve toward digital transformation, many are transitioning from manual data management to automated systems to enhance accuracy, streamline processes, and enable more robust data analysis.

However, in certain contexts, manual data entry remains relevant, such as in situations where data security and privacy concerns necessitate a more controlled, hands-on approach, or when digital solutions are not immediately available or feasible, such as when the data format is incompatible with the current setup of the digital systems in use. Another example is patient observations in rare medical conditions.

Data Governance

Data governance is a comprehensive framework and set of practices aimed at ensuring the quality, integrity, security, and responsible management of an organisation's data throughout its lifecycle. It involves the establishment of policies, procedures, and standards that guide how data is collected, stored, processed, and utilised across the organisation.

The primary goals of data governance are to enhance data reliability, promote data-driven decision-making, mitigate risks associated with data misuse, and ensure compliance with industry standards and regulations. Successful data governance initiatives often involve collaboration among different stakeholders, including data stewards, IT professionals, executives, and compliance officers, to create a unified and structured approach to managing and leveraging data as a valuable organisational asset.

The role of the finance profession in data governance is pivotal, as financial data is both sensitive and central to organisational decision-making. Finance professionals play a crucial role in defining and implementing data governance policies and procedures related to financial information, including data security. They ensure that data quality standards are upheld, contributing to the accuracy and reliability of financial reports and analyses. Finance professionals collaborate with data stewards and IT teams to establish robust metadata management practices, ensuring a clear understanding of the lineage, definitions, and usage of financial data elements.

Data Ingestion

Data ingestion is the process of collecting, importing, and processing raw data from various sources into a system or storage infrastructure for further analysis, storage, or transformation. This is a critical step in the data pipeline that facilitates the use of diverse datasets for business intelligence, analytics, and other data-driven applications. The data sources can include databases, log files, sensors, APIs, and external data streams.

Key aspects of data ingestion include:

1. Collection – Raw data is collected from different sources, which may be structured or unstructured.

2. Transport – The collected data is transported to a centralised location or data storage system. This transportation may involve data movement across networks or file systems.

3. Processing – Once the data is transported, it may undergo initial processing steps, such as cleaning, validation, or transformation, to make it suitable for the intended use.

4. Storage – The ingested data is stored in a data repository or database where it can be easily analysed.

Data ingestion is a fundamental step in the data lifecycle, enabling organisations to harness the value of their data by making it accessible and actionable for decision-making and analytics. Various tools and platforms are commonly used to facilitate efficient and scalable data ingestion processes.

Data ingestion patterns

Data ingestion patterns refer to the strategies and approaches used to collect, import, and process data from diverse sources into a system or storage infrastructure. These patterns are designed to address specific requirements, such as scalability, reliability, and real-time processing. Several common data ingestion patterns are employed based on the nature of the data and the needs of the application. The most common patterns are batch ingestion and real-time ingestion.

Batch ingestion is a data processing approach that involves collecting, processing, and importing data in predefined intervals or batches. In this pattern, data is gathered over a set period, such as hours or days, and then processed as a group. For example, ingesting sales data after the end-of-day process is complete is common.

Batch ingestion is well-suited for scenarios where the arrival time of data is not critical, and processing can be performed periodically. This approach is particularly advantageous when dealing with large volumes of data that can be efficiently processed in scheduled intervals. While it may not provide real-time insights, batch ingestion remains a reliable and scalable method for processing data in scenarios where immediate data availability is not a primary requirement.

Real-time ingestion is a data processing approach that involves the immediate and continuous collection, processing, and importing of data as it is generated. In this pattern, data is ingested and processed in near real-time (within seconds), providing the advantage of low-latency insights and quick responsiveness to changing data.

Real-time ingestion is crucial for applications where up-to-the-moment data is essential, such as real-time analytics, monitoring systems, and decision-making processes that require immediate feedback. This approach is particularly valuable in dynamic environments where timely and accurate data is a critical factor in driving informed and immediate actions. Real-time ingestion is fundamental to the agility and responsiveness required in various industries, including finance, e-commerce, and IoT applications.

Destination of ingested data

The destination of ingested data refers to the location or system where the processed and transformed data is stored for further analysis, retrieval, or utilisation. The choice of destination depends on the specific requirements of the data pipeline and the nature of the application. Several common destinations for ingested data include:

1. Data lakes – These provide a flexible storage solution for organisations dealing with vast amounts of raw and unstructured data. Platforms like Amazon S3 or Azure Data Lake Storage are commonly used for this purpose. This can be either provided by a cloud service provider or hosted internally by the organisation. The data lake includes computer storage, networking, and connectivity.

2. Cloud storage – In cloud-based environments, often provided by cloud service providers, ingested data might be stored in dedicated cloud storage services such as Amazon S3, Google Cloud Storage, or Azure Blob Storage. Cloud service providers have farms of computer storage units or data centres for use by their

customers at a fee which is more cost-effective than hosting their own infrastructure.

3. Data warehouses – Ingested data is often stored in data warehouses, which are centralised repositories designed for large-scale analytics and reporting. Popular data warehouses include Amazon Redshift, Google BigQuery, and Snowflake. Data warehouses are places within the cloud storage or data lakes to store, process, and access data.

4. Databases – In some cases, ingested data is stored in traditional relational databases, depending on the structure and complexity of the data.

The destination of ingested data is a crucial consideration in designing a data architecture, and it is often determined by factors such as data storage requirements, accessibility, retrieval speed, and the overall analytical needs of the organisation.

Data Transformation

Data transformation is the process of converting raw data from its original format into a more suitable format for analysis, reporting, or other business purposes. This crucial step in the data processing pipeline involves cleaning, enriching, and restructuring data to make it more usable and meaningful. Data transformation can encompass various tasks, depending on the specific goals of the analysis or application.

The following are the key concepts about data transformation that finance professionals should have a good understanding of.

Data lineage

Data lineage is a critical aspect of data management and analytics, providing a comprehensive record of the journey that data takes from its origin to its destination. It involves tracking the flow and transformations applied to data as it moves through different stages of a system or data pipeline.

In essence, data lineage offers a detailed map that outlines the relationships, dependencies, and processing steps associated with each data element. This transparency is invaluable for organisations seeking to understand and manage their data effectively. With data lineage, stakeholders can trace back to the source of any data point, comprehend the series of transformations it undergoes, and identify where and how it is utilised across various applications or analytical processes.

One of the key advantages of implementing data lineage is its impact on data quality, governance, and compliance. By documenting the entire lifecycle of data, organisations can ensure data accuracy and reliability. This is particularly crucial for regulatory compliance, as data lineage provides a clear audit trail that showcases how data is handled, processed, and utilised within the organisation. Moreover, data lineage supports effective data governance practices by promoting accountability, aiding in issue resolution, and enabling quick identification of the root causes of data discrepancies.

A notable example of the complications arising from a lack of data lineage can be found in the 2008 financial crisis. In the lead-up to the crisis, financial institutions extensively dealt with complex financial products, such as mortgage-backed securities (MBS) and collateralised debt obligations (CDOs), without having a clear and transparent understanding of the underlying assets' quality and origin. The absence of clear data lineage in these financial products meant that when the housing market began to falter, it became nearly impossible to trace back the mortgages and loans' quality or performance embedded in these securities. This lack of transparency and understanding significantly contributed to the financial system's instability, as confidence in these products plummeted, leading to a widespread credit freeze and, ultimately, the global financial meltdown.

As organisations increasingly rely on complex data ecosystems, the implementation of data lineage becomes instrumental in maintaining data integrity, fostering trust in analytics, and supporting informed decision-making processes.

Data quality

Data quality (DQ) is a critical aspect of any data-driven organisation, encompassing the accuracy, completeness, consistency, and reliability of the data maintained within its systems. High-quality data is essential for making informed business decisions, ensuring the success of analytics initiatives, and fostering trust among stakeholders.

Examples of common data quality issues include misspelled names, wrong addresses, missing digits in dates of birth, and wrong honorifics.

Achieving and maintaining data quality involves various processes, including data cleansing, validation, and enrichment. Data cleansing addresses issues such as missing values, inaccuracies, and inconsistencies, ensuring that the data accurately reflects the real-world entities it represents. Validation processes assess data against predefined rules and standards, identifying and rectifying anomalies. Additionally, data enrichment involves augmenting existing datasets with additional information, enhancing their value and usefulness.

The impact of poor data quality can be profound, leading to misguided decision-making, operational inefficiencies, and compromised customer experiences. Organisations with inaccurate or incomplete data risk making faulty strategic choices based on flawed insights. Addressing data quality issues requires a proactive approach involving the establishment of robust data governance practices, the implementation of data quality tools and processes, and ongoing monitoring and validation efforts. By prioritising data quality, organisations can unlock the full potential of their data assets, supporting reliable analytics, regulatory compliance, and overall business success.

Figure 6 below is an example of raw ingested data to a transformed data table that is usable across the organisation.

gested data

```
ustomerID | Name | Email | Country ------
----|------------|--------------------|--
-------- 1 | John Smith | john@email.com
 USA 2 | Jane Robinson | jane@email.com |
anada 3 | Bob Johnson| bob@email.com | UK
```

Transformed data

EmployeeID	Name	Email	Country
1	John Smith	john@email.com	USA
2	Jane Robinson	jane@email.com	Canada
3	Bob Johnson	bob@email.com	UK

Figure 6

ETL vs ELT

Extract, Transform, Load (ETL) is a fundamental data integration process designed to extract data from source systems, transform it to meet specific business requirements, and load it into a target system, typically a data warehouse or database.

In the extraction phase, data is gathered from various sources, which can include databases, applications, logs, or external datasets. The extracted data then undergoes a series of transformations during the next phase, where it is cleaned, aggregated, enriched, and structured to ensure consistency and relevance. Finally, the transformed data is loaded into the target system, making it available for analytics, reporting, and business intelligence purposes.

ETL is a critical component in the data lifecycle, enabling organisations to centralise and integrate data from disparate sources, ensuring data accuracy and providing a foundation for informed decision-making. ETL processes are often automated using specialised tools, making the data integration workflow more efficient and scalable.

Extract, Load, Transform (ELT) represents a data integration approach where data extraction and loading precede the transformation phase. In ELT, raw data is initially extracted from source systems, which can include databases, applications, or logs.

This unaltered data is then loaded into the target system, often a data warehouse or a database optimised for analytical processing. The transformation phase occurs within the target system, where the raw data undergoes necessary processing and structuring.

ELT is well-suited for modern data processing environments and scalable storage solutions, leveraging the processing power of the target system to execute complex transformations on large datasets. This approach allows organisations to benefit from the parallel processing capabilities of advanced analytics platforms and efficiently handle the increasing volumes of data generated in today's data-driven landscape. The ELT methodology offers flexibility and scalability, making it a preferred choice for organisations looking to streamline their data integration processes in the era of big data and analytics.

The choice between ETL and ELT depends on various factors, including the nature of the data, the capabilities of the target system, and the specific requirements of the business's analytics and reporting processes.

Data Labelling

Data labelling is a crucial step in the process of preparing data for ML and AI applications. It involves annotating or tagging the data with relevant labels, categories, or attributes that the ML model aims to learn and recognise. The labelled data serves as a training set for supervised learning algorithms, where the model learns to associate input features with corresponding output labels.

Data labelling can take various forms depending on the nature of the task, such as image classification, object detection, natural language processing, and more.

For example, in image classification, each image may be labelled with the object or category it represents; for example, a photo of a dog may be labelled as "dog", "animal", "quadruped", and "photography". In text classification, documents or sentences are labelled with their respective classes or topics. The accuracy and quality of data labelling significantly impact the performance of ML models.

Human annotators are often employed to label data manually, but in some cases, automated tools and techniques may be used. Data labelling is a labour-intensive process. Various data labelling techniques are employed depending on the type of data and the specific requirements of the ML task. Below are some common data labelling techniques.

1. Manual labelling

This is a crucial process in ML, where human annotators manually review and assign labels to raw data. This hands-on approach, if done properly with discipline and consistency, ensures a high level of accuracy and precision in the labelled dataset, making it a gold standard for training machine learning models.

Human annotators, often subject matter experts, meticulously examine the data and apply domain-specific knowledge to assign appropriate labels. While manual data labelling offers precision, it can be time-consuming and resource-intensive, especially for large datasets. The process involves a careful balance between maintaining labelling quality and managing the associated costs. Despite its challenges, manual data labelling remains indispensable for tasks where the data's complexity, context, or subjectivity requires human judgment to produce reliable and trustworthy labelled datasets.

2. Automatic labelling

This is a data annotation technique that leverages algorithms and predefined rules to assign labels to raw data without direct human intervention. This approach is especially useful in scenarios where the labelling task can be well-defined algorithmically, and manual annotation would be impractical or time-consuming.

Rule-based labelling involves the application of predefined rules to automatically assign labels based on specific criteria or patterns within the data. Heuristic approaches use algorithms to infer labels from features or patterns present in the data. While automatic data labelling can significantly reduce the time and cost associated with data annotation, its effectiveness depends on the clarity and reliability of the rules or algorithms applied. This technique is commonly employed in tasks where the labelling criteria are objective and can be expressed in a computational form, contributing to the efficiency of machine learning model training processes.

3. Semi-automatic labelling

Semi-automatic data labelling is an approach that combines elements of both manual and automatic labelling to annotate datasets. In this method, a limited portion of the dataset is manually labelled by human annotators, while the remaining data is left unlabelled. The labelled data is then used to train an AI model, and this model is employed to make predictions on the unlabelled data. In instances where the model is uncertain or likely to make errors, the data is then selectively labelled by human annotators. This iterative training, prediction, and targeted labelling process continues, gradually improving the model's performance with minimal manual effort. Semi-supervised data labelling is particularly advantageous when obtaining a large set of labelled data is challenging or expensive, as it leverages the power of both human expertise and ML algorithms to achieve accurate and cost-effective results.

•

Choosing the appropriate data labelling technique depends on factors such as the available resources, the complexity of the task, and the quality requirements of the labelled data. Often, semi-automatic labelling is used in practice to achieve the best results. Finance professionals must understand basic principles that could be applied to data labelling to work effectively with data engineers and data scientists.

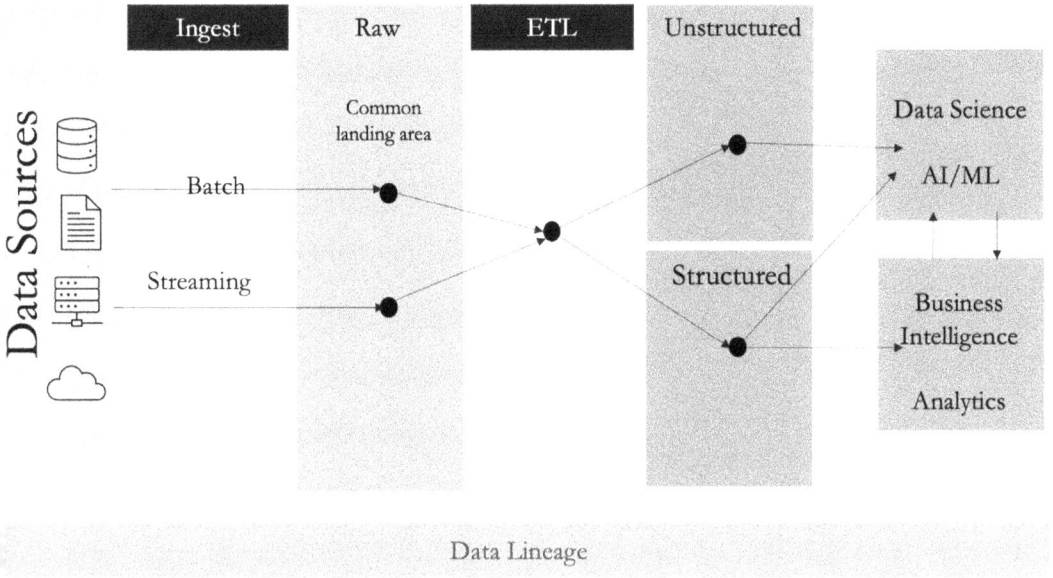

Data Sources

Ingest	Raw	ETL	Unstructured

Common landing area

Batch

Streaming

Data Science

AI/ML

Structured

Business Intelligence

Analytics

Data Lineage

Figure 7

7

AI Models

AI models are at the heart of the transformative power of machine learning (ML). These models are designed to learn and make decisions, mimicking human cognitive functions. One prominent type of AI model is the neural network, inspired by the human brain's structure. Neural networks, organised in layers of interconnected nodes, process information in a way that enables them to recognise patterns, classify data, and even generate new content. Deep learning (DL), a subset of ML, employs neural networks with multiple layers, allowing for more complex representations and abstraction of data. AI models have achieved remarkable success in various applications, from image and speech recognition to language translation and autonomous vehicles.

The effectiveness of AI models relies heavily on robust training datasets and careful fine-tuning of parameters. The ability of these models to adapt and generalise their knowledge to new, unseen data is a key factor in their utility. As AI continues to advance, researchers and developers are exploring innovative architectures and techniques to enhance the efficiency, interpretability, and ethical implications of AI models. The ongoing evolution of AI models holds promise for solving complex problems and shaping the future of technology across diverse domains.

Finance professionals must grasp the basics of AI models to stay relevant and thrive in an increasingly digitised and data-driven industry. The integration of AI in finance has become ubiquitous, offering unprecedented opportunities for analysis, prediction, and decision-making. Understanding AI models enables finance professionals to leverage advanced algorithms for risk management, fraud detection, portfolio optimisation, auditing, and market trend analysis.

Moreover, as AI technologies continue to evolve, finance professionals must be equipped to interpret and validate the outputs of these models, ensuring the ethical and responsible use of AI in financial decision-making. A solid understanding of AI models empowers finance professionals to navigate the complex landscape of financial technology, enhancing their ability to make informed decisions, adapt to changing market dynamics, and drive innovation in the financial sector. AI literacy is not just a skill, it is a necessity for finance professionals looking to harness the full potential of technology in shaping the future of finance.

Common Modelling Techniques

In the context of data science and ML, various modelling techniques are employed. Below are some common techniques:

1. Linear regression – Used for predicting a continuous variable based on one or more independent variables.

2. Decision trees – Tree-like models that make decisions based on input features. They are used for both classification and regression tasks.

3. Random forest – This combines the output of multiple decision trees to reach a single result. Its ease of use and flexibility have fuelled its adoption, as it handles both classification and regression problems.

4. Support Vector Machines (SVM) – Used for classification and regression tasks by finding a hyperplane that best separates or approximates the data.

5. Neural networks – DL models that are composed of layers of interconnected nodes, suitable for complex tasks like image recognition, natural language processing, and more.

6. K-Nearest Neighbours (KNN) – A simple algorithm that classifies a data point based on the majority class of its nearest neighbours.

7. Clustering algorithms – An algorithm such as K-Means or Hierarchical Clustering that groups similar data points.

8. Principal Component Analysis (PCA) – A dimensionality reduction technique that transforms high-dimensional data into a lower-dimensional form while retaining important information.

9. Gradient boosting – An ensemble technique that combines weak models sequentially, with each model correcting errors made by the previous ones.

10. Recurrent Neural Networks (RNN) and Long Short-Term Memory (LSTM) – These are specialised neural network architectures for handling sequential data, often used in natural language processing tasks.

The choice of modelling technique depends on the nature of the data, the problem at hand, and the desired outcome. It is common to experiment with multiple techniques to find the one that best suits a particular scenario.

Techniques Used in Text Generation

ChatGPT employs a sophisticated language model based on the Generative Pre-trained Transformer (GPT) architecture. It generates the next best word based on the pre-trained data. GPT models are designed for natural language understanding and generation. The primary technique employed in GPT is unsupervised learning from large datasets, allowing the model to learn the patterns and structures of language.

The transformer architecture, a key component of GPT, enables the model to process input data in parallel, making it highly efficient for handling sequences like sentences. Attention mechanisms within the transformer facilitate the model's focus on relevant parts of the input, allowing it to capture long-range dependencies and context.

To train ChatGPT, a diverse range of text data is used, exposing the model to a wide range of language styles and topics. The model is pre-trained on this massive dataset, learning to predict the next word in a sentence. This pre-training helps the model acquire a broad understanding of language, and then fine-tuning is performed on more specific datasets to tailor the model for conversational interactions.

The result is a language model that excels in generating coherent and contextually relevant responses based on the input it receives. It leverages a combination of pre-training and fine-tuning techniques, making it a versatile tool for various natural language processing tasks, including chat-based conversations.

Techniques Used in Image Generation

Image generation techniques involve various approaches, and different models may use a combination of methods. Below are some prominent techniques used in image generation:

1. Generative Adversarial Networks (GANs) – GANs consist of a generator and a discriminator network that are trained simultaneously. The generator creates images, and the discriminator evaluates whether they are real or generated. This adversarial process leads to the generation of increasingly realistic images.

2. Variational Autoencoders (VAEs) – VAEs focus on learning a probabilistic latent space representation of images. They can generate new images by sampling from this learned distribution. VAEs are known for their ability to generate diverse outputs and handle variations in the generated content.

3. Conditional generation – Models can be trained to generate images based on specific conditions or inputs. This involves providing additional information, such as

class labels or textual descriptions, to guide the image generation process. Conditional GANs and VAEs are common in this context.

4. Autoencoders – Autoencoders consist of an encoder and a decoder. The encoder compresses input images into a latent space representation, and the decoder reconstructs the images from this representation. Variations like denoising autoencoders can be used for generating new images by perturbing the latent space.

5. Transfer learning – Pre-trained models, often trained on large datasets for classification tasks, can be fine-tuned for image generation tasks. This allows leveraging the learned features and representations for generating new images.

6. Attention mechanisms – Inspired by the success of attention mechanisms in natural language processing, some image generation models incorporate attention mechanisms to focus on specific regions of an image during the generation process. This enhances the model's ability to capture intricate details.

7. Recurrent Neural Networks (RNNs) and transformers – While traditionally used for sequence data, RNNs and transformer architectures can be adapted for image generation tasks. For example, transformers can capture long-range dependencies and spatial relationships in images, making them effective for tasks like image captioning and generation.

These techniques are often used in combination, and advancements in image generation continue to emerge as researchers explore innovative approaches and model architectures.

Transformer Model

The transformer model is a type of DL architecture that has proven highly effective for various natural language processing

(NLP) tasks. This has given rise to the recent surgency of AI applications such as ChatGPT, Dall-E, and Bard. It was introduced by Vaswani et al. in the paper "Attention is All You Need" in 2017. The transformer architecture revolutionised natural language processing by dispensing with recurrent or convolutional layers and relying entirely on self-attention mechanisms.

Key components of the transformer model include:

1. Self-attention mechanism – Transformers use self-attention to weigh the importance of different words in a sequence in relation to each other. This allows the model to consider the entire context of a sequence in parallel, making it more efficient in capturing long-range dependencies compared to traditional sequential models.

2. Encoder-decoder architecture – Transformers are commonly employed in a sequence-to-sequence setup with an encoder-decoder structure. The encoder processes the input sequence, and the decoder generates the output sequence. This architecture is versatile and has been used in tasks like machine translation and text summarisation.

3. Multi-head attention – To enhance the model's ability to focus on different aspects of the input, self-attention is often implemented with multiple attention heads. Each head learns a different attention pattern, and their outputs are concatenated or linearly combined to capture diverse information.

4. Positional encoding – Since transformers do not inherently understand the sequential order of data, positional encoding is added to the input embeddings to provide information about the position of each token in the sequence.

5. Feedforward neural networks – Transformers incorporate feedforward neural networks after the self-

attention mechanism in both the encoder and decoder. These networks contribute to capturing non-linear relationships in the data.

6. Layer normalisation and residual connections: To aid training and mitigate the vanishing/exploding gradient problem, each sub-layer (like self-attention or feedforward) in a transformer block is followed by layer normalisation and a residual connection.

Transformers have demonstrated state-of-the-art performance in various natural language processing tasks, such as machine translation, language modelling, and text generation. Moreover, the architecture's attention mechanism has found applications beyond NLP, including computer vision tasks like image classification. The success of transformers has led to their widespread adoption in many areas of AI.

Attention Mechanism

Imagine you are reading a story, and certain parts of the story catch your attention more than others. For instance, when a character is introduced or a surprising event occurs, you naturally focus on those details. The attention mechanism in a computer program works in a similar way.

The attention mechanism is like a smart reader for computers. When the computer processes information, it doesn't treat all parts equally. Instead, it pays more attention to important bits, just like you focus on key elements in a story. It assigns different levels of importance to different parts of the information.

This attention mechanism helps the computer in tasks like language translation or summarising text as it allows the program to understand the context and relationships between different parts of the information, making it smarter and more effective in various tasks.

More technically, the attention mechanism is a crucial component in DL architectures, originally popularised by its inclusion in the transformer model for natural language processing. It enables models to selectively focus on different parts of the input sequence when making predictions or generating output. The attention mechanism is particularly powerful in capturing long-range dependencies and improving the model's performance on tasks that involve sequential or structured data.

The following is a simplified explanation of how the attention mechanism works:

1. Key, query, and value – In the context of attention, the input sequence is typically represented as key-value pairs. For each element in the sequence, there are associated key, query, and value vectors. The query vector is used to compare against the key vectors to determine the importance (attention weight) of each element.

2. Attention scores – The attention mechanism computes attention scores by measuring the similarity between the query vector and the key vectors. This is often done using the dot product, but other methods like scaled dot-product attention are also common. The higher the attention score, the more focus is given to the corresponding value vector.

3. Weighted sum – The attention scores are normalised using a softmax function to obtain weights that sum to 1. These weights are then used to compute a weighted sum of the value vectors. This weighted sum, known as the attention output, represents the information attended to by the model.

4. Multi-head attention – To capture different aspects and patterns in the data, the attention mechanism is often applied with multiple sets of key, query, and value projections, known as attention heads. The outputs from these attention heads are then concatenated or linearly combined to form the final output.

5. Positional encoding – For sequential data, like in natural language processing, positional encoding is added to the input embeddings to provide information about the order or position of elements in the sequence. This helps the attention mechanism understand the sequential context.

Consider the below practical example of an attention mechanism using a sentence.

Input Sentence: "The cat sat on the mat."

1. Key, query, and value: Each word in the sentence has associated key, query, and value vectors. For simplicity, let's use the same vectors for all three.

Word	Key	Query	Value
The	K_1	Q_1	V_1
cat	K_2	Q_2	V_2
sat	K_3	Q_3	V_3
on	K_4	Q_4	V_4
the	K_5	Q_5	V_5
mat	K_6	Q_6	V_6

Attention scores: Let's say we want to focus on the word "cat" in the output. The algorithm selects "cat" as it can differentiate the subject, object, and verb in the sentence. The attention scores are calculated by comparing the query vector for "cat" (Q_2) with the key vectors for all words.

Word	Attention Score
The	$A_1 = \text{softmax}(Q_1 * K_2)$
cat	$A_2 = \text{softmax}(Q_2 * K_2)$
sat	$A_3 = \text{softmax}(Q_3 * K_2)$

on	$A4 = \text{softmax}(Q4 * K2)$
the	$A5 = \text{softmax}(Q5 * K2)$
mat	$A6 = \text{softmax}(Q6 * K2)$

Weighted sum: The attention scores are normalised using a softmax function, producing weights that sum to 1. These weights are then used to compute a weighted sum of the value vectors.

Weighted Sum = $A1 * V1 + A2 * V2 + A3 * V3 + A4 * V4 + A5 * V5 + A6 * V6$

In this example, the attention mechanism allows the model to focus more on the word "cat" when generating an output, considering its importance in the context of the sentence. This mechanism enables the model to dynamically adjust its attention based on the specific information it needs during the generation process.

Multimodals

In the context of AI, "multimodal" refers to systems or models that can understand and generate information from multiple modalities or types of data. Modalities can include text, images, audio, video, and more. Multimodal AI aims to integrate and process information from different sources to achieve a more comprehensive understanding of the input data.

For example, a multimodal AI model could be capable of:

1. Image captioning – Generating textual descriptions of the content in images.

2. Visual Question Answering (VQA) – Answering questions about the content of images.

3. Speech-to-text with image context – Transcribing spoken words into text while considering accompanying visual information.

4. Language translation with image context – Translating text from one language to another while taking into account relevant visual data.

These applications highlight the versatility of multimodal AI, which can enhance the capabilities of systems to perform tasks that involve different types of sensory input. The goal is to create AI models that can understand and generate information in a way that more closely resembles human abilities, where we seamlessly integrate information from various senses. This will further enable the growth and maturity of AI in the future. An example of a multimodal AI is Google's Gemini.

Modelling Languages

Several programming languages are commonly used in data science due to their rich ecosystems and libraries tailored for tasks like data analysis, ML, and statistical modelling. Below are some of the prominent programming languages in data science.

1. Python

Python is a versatile and dynamic programming language known for its readability, simplicity, and extensive ecosystem. Python has become a cornerstone in various domains, including web development, automation, scientific computing, and particularly in data science and machine learning, since its development. Its clean syntax and readability make it an ideal choice for beginners and experienced developers.

Python's appeal lies in its rich set of libraries, such as NumPy, pandas, and scikit-learn, which facilitate tasks like data manipulation, analysis, and machine learning. Additionally, its community-driven development model has led to a vast repository of third-party packages, fostering innovation and collaboration. Python's widespread adoption in diverse industries and its emphasis on simplicity and efficiency solidifies its position as one of the most popular and accessible programming languages in today's technology landscape.

The below script calculates and prints the answer for 1 + 1:

```
# Calculate the sum of 1 + 1

result = 1 + 1

# Print the answer in the format "1 + 1 = "

print(f"1 + 1 = {result}")
```

2. R

R is a powerful and open-source programming language designed specifically for statistical computing and data analysis. Similar to Python, R has become a staple in academia and industries that heavily rely on statistical modelling and data visualisation.

One of R's distinguishing features is its comprehensive collection of packages, including ggplot2 for expressive data visualisation and dplyr for efficient data manipulation. R's syntax is particularly well-suited for statistical tasks, making it a preferred choice among statisticians, data scientists, and researchers. The language's vibrant community actively contributes to its extensive ecosystem, ensuring that R remains at the forefront of advancements in statistical computing. With its emphasis on statistical techniques, data exploration, and visualisation, R continues to play a crucial role in the data science landscape, offering a specialised and powerful toolset for those engaged in quantitative analysis and research.

The below script calculates and prints the answer for 1 + 1:

```
# Calculate the sum of 1 + 1

result <- 1 + 1

# Print the result

cat("1 + 1 =", result, "\n")
```

3. *Julia*

Julia is a high-performance and dynamic programming language designed for technical and scientific computing. Developed with a focus on combining the speed of low-level languages like C and Fortran with the expressiveness of high-level languages like Python and R, Julia aims to address the challenges of numerical and computational tasks.

Julia's syntax is both simple and familiar, making it accessible to a broad audience. Julia's key strength lies in its Just-In-Time (JIT) compilation, allowing for near-native performance and efficient execution of numerical algorithms. With growing popularity in fields like ML, data science, and simulation-based research, Julia serves as a powerful tool for those seeking a high-performance language that seamlessly balances speed and ease of use in scientific computing applications.

4. *Scala*

Scala, a versatile and statically typed programming language, seamlessly blends functional and object-oriented programming paradigms. The primary goal of Scala is to provide a concise and expressive language that runs on the Java Virtual Machine (JVM). Scala's syntax is concise, and its compatibility with Java allows for smooth integration with existing Java codebases. It offers advanced features like pattern matching, immutability by default, and a strong type of system, making it an ideal choice for large-scale, concurrent, and distributed systems.

Scala's flexibility, along with its compatibility with popular frameworks like Apache Spark, has positioned it as a language of choice for applications demanding both expressiveness and scalability. The language's emphasis on functional programming constructs and support for concurrent and parallel programming contribute to its appeal in modern software development, particularly in the context of data-intensive and distributed computing environments.

5. SAS

SAS (Statistical Analysis System) is a software suite developed by SAS Institute for advanced analytics, business intelligence, and data management. SAS is widely used across industries for statistical analysis, data exploration, and predictive modelling. Known for its robust capabilities in handling and analysing large datasets, SAS offers a comprehensive range of modules and solutions for various tasks, including data mining, machine learning, and statistical modelling.

Professionals in fields such as finance, healthcare, and market research often rely on SAS for its reliability, scalability, and ability to process diverse data types. With a strong emphasis on data integrity, SAS provides a versatile platform for extracting actionable insights from complex datasets, contributing to its long-standing reputation as a leader in the field of analytics and business intelligence.

Python and R are among the most widely adopted languages in the data science community due to their extensive libraries and active communities. Many data scientists also use a combination of languages based on the demands of a particular project. There are many proprietary and open-source programs used in different data modelling applications.

Modelling Tools

There are various tools and frameworks available for building ML models. Most of these tools bring programming languages, data storage, and best practices together to simplify model building. Below are some popular ones:

1. TensorFlow – Developed by Google, TensorFlow is an open-source framework widely used for building and deploying ML models. It offers a comprehensive set of tools for various tasks, including DL.

2. PyTorch – Developed by Facebook's AI Research lab (FAIR), PyTorch is another popular open-source deep learning framework. It is known for its dynamic

computational graph, making it more intuitive for researchers and developers.

3. Scikit-learn – A simple and efficient tool for data analysis and modelling, scikit-learn is built on NumPy, SciPy, and Matplotlib. It provides simple and efficient tools for data mining and data analysis, accessible to everybody.

4. Keras – Keras is an open-source DL Application Programming Interface (API) written in Python. It serves as a high-level interface for neural networks and runs on top of other popular deep learning frameworks like TensorFlow and Theano.

5. Jupyter Notebooks – Jupyter is an open-source web application that allows you to create and share documents that contain live code, equations, visualisations, and narrative text. It is widely used in data science and ML for interactive development.

6. Apache Spark MLlib – For big data processing and ML, Apache Spark MLlib is a scalable ML library. It integrates with the Apache Spark framework, making it suitable for large-scale data processing.

7. Caffe – A DL framework developed by the Berkeley Vision and Learning Centre (BVLC) for image classification, segmentation, and more.

8. MXNet – An open-source DL framework that is particularly known for its flexibility and efficiency, supporting both symbolic and imperative programming.

9. Microsoft Azure Machine Learning – This is a cloud-based service for building, training, and deploying ML models. It provides a variety of tools and services to support the end-to-end ML lifecycle.

10. H2O.ai – This offers an open-source platform, designed for scalable and distributed ML. It supports various algorithms and integrates well with popular programming languages.

The choice of a tool often depends on factors such as the specific use case, familiarity, community support, and scalability requirements. Many practitioners use a combination of these tools based on their needs. While the above are tools used in practice, finance professionals are not required to understand how and when to use them.

Nonetheless, a general understanding of technologies and processes is required to be effective in the modern workplace. Organisations not only expect finance professionals to work closely with domain experts, but they also have to make strategic decisions in the organisation that leverage AI technologies.

AI Model Training and Continuous Learning

AI model training is a critical phase in the development of AI systems, involving the iterative process of refining a model's parameters to enhance its ability to make accurate predictions or decisions. At the core of this process is the utilisation of labelled datasets, where the model learns patterns and relationships between input features and target outputs.

The data is typically split into training and validation sets, with the former used to update the model's parameters and the latter employed to evaluate its performance. The choice of an appropriate model architecture is crucial, with options ranging from traditional ML algorithms to sophisticated DL architectures. During training, the model undergoes multiple iterations, adjusting its internal parameters through optimisation. This continuous refinement is essential to ensure the model's accuracy and effectiveness in solving the specified problem.

The success of AI model training also depends on careful consideration of various factors, such as hyperparameter tuning, regularisation techniques, and the choice of evaluation metrics. Hyperparameters, like learning rates and batch sizes, influence the model's learning process and generalisation performance. Regularisation methods help prevent overfitting by penalising overly complex models. Evaluation metrics, such as accuracy or precision, guide the assessment of the model's performance and suitability for the intended task.

Additionally, ethical considerations and transparency in the training process are gaining increasing importance, as they address concerns related to bias, fairness, and the responsible use of AI technology. As AI continues to advance, ongoing research and development in model training methodologies play a pivotal role in ensuring that AI systems not only perform effectively but also align with ethical standards and societal expectations.

AI modelling example on accounting

Creating an AI model for a double-entry system in accounting involves training a model to understand and process financial transactions, ensuring that debits and credits are accurately recorded to maintain accounting balance.

Task: Automated double-entry accounting.

Data: A dataset of financial transactions, each with details like transaction type, accounts involved, and amounts. Each transaction is labelled with the correct debit and credit entries.

Data preparation: Format the financial data, ensuring consistency in transaction details, account categories, and amounts.

Model architecture: Develop a neural network, perhaps a combination of recurrent and dense layers, designed to analyse transaction details and predict appropriate debit and credit entries.

Training process using supervised learning: Feed the model batches of labelled transactions. The model learns to associate transaction details with correct debit and credit entries, adjusting its parameters during training.

Evaluation: Assess the model's performance on a validation set of transactions it hasn't seen before. Metrics such as accuracy and precision are used to measure the model's ability to correctly predict double-entry records.

Fine-tuning: If the model struggles with certain transaction types or patterns, then fine-tune its parameters or architecture to improve accuracy.

Testing: Once satisfied with the model's performance, test it on new, unseen financial transactions to ensure its real-world effectiveness.

Use case: The trained model can then be integrated into accounting software, automatically processing transactions and generating accurate double-entry records. This helps reduce manual effort and minimises the risk of human error in accounting processes.

AI modelling example on auditing

Creating an AI model for auditing involves training a system to analyse financial data and identify anomalies or patterns that might indicate irregularities or potential issues.

Task: Automated anomaly detection in financial auditing.

Data: A dataset of financial transactions, including details such as transaction type, amounts, dates, and associated accounts. Data is labelled indicating whether each transaction is normal or potentially problematic based on historical audit findings.

Data preparation: Format the financial data, ensuring consistency in transaction details and labels indicating normal or anomalous behaviour.

Model architecture: Design a ML model, possibly a combination of anomaly detection algorithms and neural networks, to learn patterns in normal transactions and identify deviations.

Feature engineering: Extract relevant features from the data, such as transaction frequency, amounts, and historical trends.

Training process using supervised learning: Train the model on labelled data, exposing it to both normal and anomalous transactions. The model learns to distinguish between the two based on the features provided.

Evaluation: Assess the model's performance on a validation set, measuring metrics like precision and recall to ensure it accurately identifies anomalies without too many false positives.

Fine-tuning: Adjust model parameters or consider incorporating additional features to enhance its ability to detect anomalies effectively.

Testing: Once satisfied with the model's performance, test it on new financial data to evaluate its real-world effectiveness.

Use case: The trained model can then be employed in an auditing system to automatically analyse large volumes of financial transactions. It helps auditors focus their attention on potentially problematic areas, streamlining the auditing process and improving efficiency.

AI modelling example on tax

Building an AI model for company tax involves training a system to analyse financial data, understand tax regulations, and assist in tax compliance.

Task: Automated tax compliance for companies.

Data: A dataset of company financial records, including income statements, balance sheets, and other relevant financial details such as historical tax filings labelled with correct tax amounts.

Data preparation: Format financial data, ensuring consistency in accounting practices and labelling each entry with the correct tax classification.

Model architecture: Design a ML model, possibly a regression model or ensemble of models, to predict the appropriate tax amount based on financial data.

Feature selection: Identify relevant features from the financial data, such as revenue, expenses, profit margins, and other factors influencing tax liability.

Training process using supervised learning: Train the model on historical data, allowing it to learn the relationships between financial variables and tax liabilities. Adjust model parameters to minimise the difference between predicted and actual tax amounts.

Evaluation: Assess the model's performance on a validation set using metrics like Mean Absolute Error or R-squared to ensure accurate predictions.

Fine-tuning: Fine-tune the model based on feedback from tax experts or additional insights into company-specific tax considerations.

Testing: Once satisfied with the model's accuracy, test it on new financial data to evaluate its real-world effectiveness in predicting tax liabilities.

Use Case: The trained model can be integrated into tax software or financial systems, providing companies with an automated tool for predicting tax amounts and ensuring compliance with tax regulations.

As with any new software platforms, it is important to note that the actual implementation of such a model would require collaboration with tax experts, consideration of specific tax codes and regulations, and ongoing monitoring and updating to adapt to changes in tax laws or company financial practices.

AI Model Evaluation

AI model evaluation is a crucial step to ensure the effectiveness and reliability of a trained model. The evaluation process involves assessing how well the model performs on data it hasn't seen before, providing insights into its generalisation capabilities. Below are key aspects of AI model evaluation.

1. Splitting the data

Training set: The portion of the dataset used to train the model.

Validation set: A separate subset used during training to tune hyperparameters and avoid overfitting.

Test set: Data not seen by the model during training or validation, reserved for final evaluation.

2. Metrics

Accuracy: The proportion of correctly classified instances, common for classification tasks.

Precision and recall: This is particularly important for imbalanced datasets. Precision focuses on the accuracy of positive predictions, while recall measures the ability to capture all positive instances.

F1 Score: A balance between precision and recall, useful when an uneven class distribution occurs.

Mean Absolute Error (MAE), Mean Squared Error (MSE), or R-squared: Common for regression tasks.

3. Confusion matrix

This is a table that visualises the model's performance, showing true positives, true negatives, false positives, and false negatives.

4. ROC curve (Receiver Operating Characteristic)

This is especially relevant for binary classification problems, the ROC curve plots the true positive rate against the false positive rate at various threshold settings.

5. Learning curves

These are plots that illustrate how the model's performance changes with respect to the amount of training data.

6. Cross-validation

This process repeatedly splits the data into training and validation sets, then trains and evaluates the model multiple times to get a more robust performance estimate.

7. Interpreting results

Consider the context of the specific task and understand the implications of the chosen metrics. High accuracy doesn't necessarily mean the model is perfect, especially if the dataset is imbalanced.

8. Domain expert feedback

Collaborate with domain experts to validate the model's outputs, ensuring it aligns with real-world expectations and requirements.

9. Bias and fairness evaluation

Assess the model's performance across different demographic groups to identify and mitigate potential biases.

Continuous monitoring and periodic re-evaluation are essential, especially as the model encounters new data or faces changes in the underlying distribution. This iterative process helps maintain the model's relevance and reliability over time.

AI Model Deployment

AI model deployment involves making a trained ML model accessible for use in real-world applications. Below are key steps and considerations for deploying an AI model:

1. Preprocessing and data compatibility – This ensures that the input data format aligns with what the model expects. Implement any necessary preprocessing steps to handle incoming data appropriately.

2. Scalability and performance – Optimise the model for performance, considering factors like response

time and resource utilisation. This may involve model quantisation, pruning, or other techniques to make the model suitable for deployment on various platforms.

3. Containerisation – Package the model, along with its dependencies, into a container (e.g., Docker). Containerisation ensures consistent behaviour across different environments and facilitates easy deployment.

4. API development – Create an API that allows other software applications to interact with and make predictions using the model. Common API protocols include RESTful APIs or gRPC.

5. Security measures – Implement security measures to protect both the model and the data it processes. This includes encryption, access controls, and secure communication channels.

6. Monitoring and logging – Set up monitoring systems to track the model's performance, detect anomalies, and log relevant information. Monitoring ensures that the deployed model continues to operate effectively over time.

7. Versioning – Implement version control for your models to keep track of changes and updates. This is crucial for maintaining and rolling back to previous versions if needed.

8. Deployment environment – Choose an appropriate deployment environment, whether it is on-premises, cloud-based (e.g., AWS, Azure, Google Cloud), or edge computing devices, based on the specific requirements of your application.

9. Continuous integration/Continuous deployment (CI/CD) – Implement CI/CD pipelines to automate the testing and deployment processes, ensuring smooth updates and minimising downtime.

10. Documentation – Create comprehensive documentation for the deployed model, including information on how to use the API, expected input formats, and any specific requirements.

11. A/B testing – If applicable, consider implementing A/B testing to compare the performance of different model versions in a production environment.

12. Feedback loop – Establish a feedback loop to collect insights from the deployed model's performance in real-world scenarios. This information can be valuable for future model improvements.

13. Legal and privacy compliance – Ensure that the deployment complies with legal and ethical considerations, especially concerning data privacy and fairness.

14. User training and support – Provide training and support resources for users who will be interacting with the deployed model, especially if it involves non-technical stakeholders.

Deploying an AI model is a complex process that requires collaboration between data scientists, software engineers, and domain experts to ensure a successful integration into the target environment.

8

AI Model Interface

The interface of an AI model serves as the bridge between users and the sophisticated algorithms working behind the scenes. Whether through a sleek GUI, a mobile app, or a conversational chatbot, the interface determines how users interact with the model's capabilities. For example, a user-friendly web interface might empower individuals to effortlessly utilise image recognition for various purposes, simply by uploading images and receiving instant insights. The popular chatbot ChatGPT is accessible via web and app interfaces.

On the other hand, a command-line interface might cater to developers and technical users who prefer direct interaction with the model through text-based commands, providing a more efficient and programmatic approach.

In the field of AI, interfaces extend beyond traditional visual interactions to include voice-activated interfaces, allowing users to seamlessly communicate with the model using natural language. Imagine a voice-controlled virtual assistant that understands spoken commands and responds with relevant information or performs specific tasks. These diverse interfaces cater to the varied preferences and technical proficiencies of users, contributing to the accessibility and usability of AI technologies in different domains and applications.

A few common interfacing approaches are outlined below.

Role of Libraries

Libraries play a crucial role in the development of AI interfaces by providing pre-built functionalities, tools, and frameworks that streamline the implementation process. Below are the key roles libraries play in creating effective AI interfaces.

1. Removing complexity

Libraries abstract the intricate details of AI model integration, allowing developers to focus on designing user interfaces without delving deep into the complexities of underlying algorithms or model deployment.

2. Efficient development

Ready-made libraries accelerate the development process, enabling quicker prototyping and implementation of AI interfaces. Developers can leverage existing code for common functionalities, saving time and effort.

3. Consistency and standards

Libraries often follow best practices and industry standards, ensuring that AI interfaces adhere to design and functionality conventions. This consistency contributes to a more user-friendly experience.

4. Cross-platform compatibility

Libraries designed for cross-platform development facilitate the creation of interfaces that can run seamlessly on various devices and operating systems, enhancing accessibility.

5. Integration with AI services

Libraries simplify the integration of AI models into different interfaces. For example, API development libraries like Flask or FastAPI facilitate the creation of APIs for model deployment, enabling smooth interaction between the model and various applications.

6. User interaction enhancement

Libraries for graphical interfaces, like React for web development or SwiftUI for iOS applications, provide tools for creating visually appealing and responsive interfaces, enhancing the overall user experience.

7. Scalability and maintenance

Libraries often come with features that enhance the scalability and maintainability of AI interfaces. This includes tools for version control, modular design, and updates that make it easier to manage evolving projects.

8. Community support

Widely adopted libraries benefit from active community support. Developers can leverage community-contributed resources, documentation, and forums to troubleshoot issues and stay updated on best practices.

9. Security measures

Libraries may include security features or recommendations, helping developers implement secure practices and protect AI interfaces from potential vulnerabilities.

10. Customisation and flexibility

Libraries provide a foundation for customisation, allowing developers to tailor AI interfaces to specific requirements while still benefiting from the library's core functionality.

By serving as building blocks for AI interface development, libraries empower developers to create robust, feature-rich applications that effectively integrate AI capabilities while meeting user expectations and industry standards.

Role of Software Development Kits

Software Development Kits (SDKs) play a significant role in facilitating the integration of AI models into interfaces, providing developers with tools and resources to enhance the development process. Below are the key roles that SDKs play in AI model interfaces.

1. Model integration

SDKs often include pre-built connectors and APIs that simplify the integration of AI models into various applications. This abstraction reduces the complexity of interfacing with the underlying ML algorithms.

2. Abstraction of technical details

SDKs abstract technical intricacies, allowing developers to interact with AI models through high-level functions and methods. This streamlines the implementation process, making it more accessible for a broader range of developers.

3. Platform independence

Many SDKs are designed to be platform-agnostic, enabling developers to integrate AI models into interfaces across different operating systems and devices. This versatility contributes to cross-platform compatibility.

4. Consistent user experience

SDKs often provide standardised methods for user interactions, ensuring a consistent and cohesive experience across various applications that leverage the same AI model.

5. Development acceleration

SDKs offer a set of pre-built tools and utilities that accelerate development by providing ready-made components, reducing the need for developers to start from scratch.

6. Documentation and best practices

SDKs typically come with comprehensive documentation, tutorials, and best practices, guiding developers on how to effectively use the provided tools. This support fosters efficient implementation and encourages adherence to industry standards.

7. Community and ecosystem

Active developer communities often support SDKs which fosters collaboration, knowledge sharing, and the development of additional resources, plugins, or extensions that enhance the functionality of the SDK.

8. Security features

Some SDKs include security measures and recommendations, helping developers implement secure practices when dealing with AI models and sensitive data.

9. Scalability and performance optimisation

SDKs may provide features for optimising the performance of AI models, including techniques for scaling the model to handle increased workloads efficiently.

10. Updates and Maintenance

SDKs may include mechanisms for receiving updates and patches, ensuring developers can adapt their interfaces to evolving AI models and industry standards.

11. Testing and debugging tools

SDKs often offer tools for testing and debugging AI model integrations, assisting developers in identifying and resolving issues during the development process.

In summary, SDKs serve as valuable resources that empower developers to seamlessly incorporate AI models into their interfaces. By providing abstraction layers, standardised interfaces, and a supportive ecosystem, SDKs contribute to the accessibility, efficiency, and effectiveness of AI model integration.

Role of Application Programming Interfaces

Considered a modern IT architectural component, Application Programming Interfaces (APIs) play a pivotal role in shaping the interaction between AI models and interfaces, serving as the bridge that enables seamless communication between different software components. Below are the key roles that APIs play in the context of AI model interfaces.

1. Model exposure

APIs allow AI models to be exposed and accessed by external applications or services. This facilitates the integration of AI capabilities into various interfaces, ranging from web applications to mobile apps.

2. Standardised communication

APIs define a standardised set of rules and protocols for communication between software components. This standardisation ensures consistency in how interfaces interact with AI models, promoting interoperability.

3. Input and output handling

APIs specify how to send input data to the AI model and receive output predictions. This formalised structure streamlines the process of feeding data into the model and extracting meaningful results.

4. Abstraction of model complexity

APIs abstract the underlying complexities of AI models, allowing developers to interact with them through well-defined endpoints and methods. This abstraction simplifies the integration process and makes AI more accessible.

5. Cross-platform compatibility

APIs enable AI models to be integrated into a wide range of interfaces regardless of the underlying technologies. This cross-platform compatibility is crucial for deploying AI models across diverse applications and devices.

6. Scalability

APIs facilitate the scalability of AI model interfaces by allowing multiple applications to simultaneously access and utilise the same model. This is essential for handling increased workloads or user demands.

7. Real-time interaction

APIs enable real-time interaction with AI models, providing instantaneous responses to user queries or requests. This responsiveness is particularly valuable in applications where timely predictions are essential.

8. Version control

APIs often support versioning, allowing developers to manage different versions of the AI model. This is crucial for maintaining backward compatibility and ensuring a smooth transition when updates are made to the model.

9. Security measures

APIs can include security features such as authentication and encryption, safeguarding the communication between interfaces and AI models. This is vital for protecting sensitive data and ensuring the integrity of the interactions.

10. Flexibility and customisation

APIs offer flexibility in how developers interact with AI models. They can be tailored to suit the specific needs of different interfaces, providing a customisable approach to leveraging AI capabilities.

11. Monitoring and analytics

APIs often come with tools for monitoring usage, tracking performance, and collecting analytics data. This information is valuable for assessing the effectiveness of the AI model in real-world scenarios.

In essence, APIs act as enablers, empowering developers to integrate AI models seamlessly into their interfaces while providing a standardised and efficient means of communication.

In conclusion, finance professionals should understand AI model interfaces because they directly impact their ability to leverage advanced technologies for more efficient and effective financial operations. As AI becomes increasingly integrated into financial systems, professionals in the finance sector need to comprehend how AI models are interfaced with their existing tools and workflows. It enables them to collaborate effectively with data scientists and technologists, ensuring that AI models align with financial regulations, ethical standards, and the specific needs of their organisations.

Understanding AI model interfaces allows finance professionals to harness the power of ML for tasks such as predictive analytics, risk management, fraud detection, and investment strategy optimisation. In short, by having a grasp of AI, finance professionals can make informed decisions, drive innovation, and stay competitive.

A practical application of concepts

Below is an example of an AI model for forensic analysis integrated into internal auditing for a large organisation using unsupervised learning techniques.

Task: Anomaly detection in internal auditing.

Data: A comprehensive dataset containing financial transactions, employee activities, scanned documents and system logs.

There are no labelled instances of fraudulent activities; the AI model is trained on data to detect and identify anomalies.

1. Data preparation: Collect and preprocess data, ensuring consistency and compatibility across different sources.

2. Feature engineering: Extract relevant features such as transaction amounts, timestamps, access logs, and employee interactions.

3. Model architecture: Implement an unsupervised learning algorithm for anomaly detection. Train the model on normal behaviour patterns, considering the multi-modal nature of data sources.

4. Integration with internal auditing: Develop an API that interfaces with the AI model. This API is integrated into the organisation's internal auditing system, allowing auditors to submit data for analysis and receive insights.

5. Real-time monitoring: The AI model continuously monitors incoming data in real-time, identifying anomalies or unusual patterns that deviate from established norms.

6. Alert generation: Whenever the AI model detects anomalies, it generates alerts for auditors. These alerts include details about the suspicious activity, aiding auditors in further investigation.

7. Interactive dashboard: Develop an interactive dashboard that visualises the detected anomalies, providing auditors with a user-friendly interface for exploring and understanding the flagged activities.

Use case: An auditor notices an anomaly in financial transactions, where a certain employee's access pattern and transaction behaviour deviate significantly from the norm.

The AI model's alert prompts the auditor to investigate further.

The investigation reveals a potentially fraudulent activity where the employee accessed sensitive financial data without proper authorisation.

The organisation takes corrective measures, and the AI model adapts and learns from the identified anomaly.

This AI model aids internal auditing by automating the detection of unusual patterns or anomalies that may indicate fraudulent or unauthorised activities within a large organisation. It enhances the efficiency of auditors by providing real-time insights and alerts, contributing to a proactive approach to maintaining the integrity and security of organisational processes.

9

ML Ops

ML Ops, short for Machine Learning Operations, refers to the set of practices and tools that aim to streamline and automate the end-to-end ML lifecycle. The goal of ML Ops is to enhance collaboration and communication between data scientists, engineers, and operations teams, to ensure that ML models can be deployed, monitored, and maintained in a scalable and efficient manner. ML Ops is essentially an extension of DevOps principles tailored to the unique challenges of ML.

ML Ops is crucial as it addresses the unique challenges associated with deploying, managing, and maintaining AI models at scale. ML Ops streamlines the end-to-end ML processes, fostering collaboration between data scientists, engineers, and operations teams. With automated scaling, monitoring, and logging, ML Ops enhances the reliability and performance of deployed models in production.

Furthermore, ML Ops establishes feedback loops for continuous improvement, allowing organisations to adapt and refine models based on real-world performance data. Ultimately, ML Ops is vital for organisations aiming to harness the full potential of ML in a systematic, scalable, and secure manner, enabling them to deploy and manage models effectively in dynamic and evolving business environments.

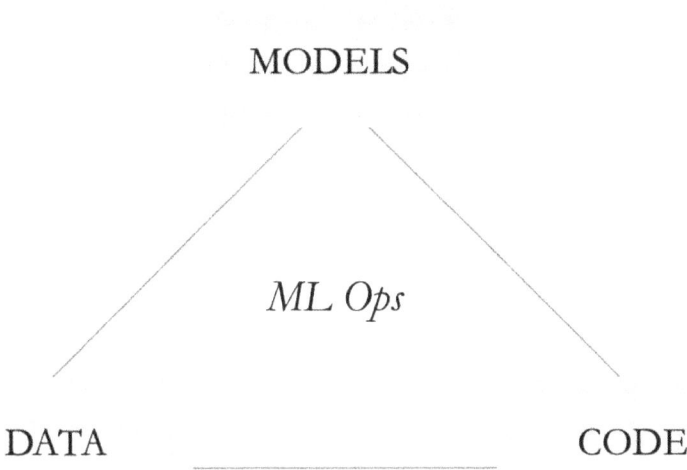

MODELS

ML Ops

DATA _____ CODE

Figure 8

Example of the Lack of ML Ops

XYZ Corporation is an e-commerce company that embarked on an ambitious project to implement AI-driven product recommendations without a well-established ML Ops function. The company's data science team developed sophisticated ML models to personalise product recommendations based on user behaviour, preferences, and historical data.

However, without proper ML Ops practices in place, the deployment of these models into the production environment became a cumbersome and error-prone process.

As the company lacked automated testing and continuous integration, deploying new versions of the recommendation models was a manual and time-consuming task. This resulted in frequent disruptions to the online shopping experience, causing frustration among users. Additionally, the absence of effective monitoring and logging mechanisms meant that issues in model performance often went unnoticed until customers reported problems.

Furthermore, the company struggled with scaling the infrastructure to handle increased user traffic during peak periods. The lack of automated scaling mechanisms led to server crashes and a degradation in the quality of recommendations during high-demand periods, impacting both user satisfaction and sales.

Ultimately, the company's failure to implement ML Ops practices resulted in wasted efforts on AI development. Despite having powerful ML models, the company faced operational challenges that hindered the effective deployment, monitoring, and scalability of its AI-driven product recommendation system.

This example highlights the importance of ML Ops in ensuring the seamless integration of ML models into production environments while avoiding disruptions and maximising the value derived from AI initiatives.

Composition of ML Ops Team

The composition of an ML Ops team can vary depending on the size and structure of the organisation, as well as the specific requirements of the projects. A typical ML Ops team often includes professionals with diverse skills and responsibilities.

1. Data scientists

- Responsibilities: Design, test, develop, deploy, and maintain ML models, algorithms, and experiments.

- Key skills: Deep understanding of ML algorithms, data science, and model training.

2. Data engineers

- Responsibilities: Build and maintain the data infrastructure, pipelines, and data extraction processes needed for model training and deployment.

- Key skills: Proficiency in data engineering, database management, and knowledge of big data technologies.

3. DevOps engineers

- Responsibilities: Implement and manage continuous integration/continuous delivery (CI/CD) pipelines, automate infrastructure provisioning, and ensure the scalability and reliability of ML systems.

- Key skills: Expertise in DevOps practices, infrastructure as code, and containerisation.

4. Software engineers

- Responsibilities: Develop and maintain software applications for integrating ML models into production systems, creating APIs, and building user interfaces for monitoring and control.

- Key skills: Proficiency in programming languages, software development methodologies, and API design.

5. Infrastructure specialists/architects

- Responsibilities: Manage and optimise the underlying infrastructure, including cloud services, servers, and networking, to support the deployment and operation of ML models.

- Key skills: Knowledge of cloud platforms (e.g., AWS, Azure, GCP), server management, and network administration.

6. QA/Test engineers

- Responsibilities: Develop and execute testing strategies for ML models and systems to ensure reliability, performance, and security.

- Key skills: Testing methodologies, automation frameworks, and a good understanding of machine learning concepts.

7. Security experts

- Responsibilities: Implement security best practices to protect data, models, and infrastructure, ensuring compliance with regulatory standards.

- Key skills: Cybersecurity knowledge, familiarity with data privacy regulations, and experience in securing ML systems.

8. Operations and support

- Responsibilities: Monitor deployed models, handle incidents, and provide support to troubleshoot and resolve issues.

- Key skills: Operations management, incident response, and a good understanding of the deployed ML systems.

9. Business analysts/Product owners

- Responsibilities: Bridge the gap between technical teams and business stakeholders, defining requirements, prioritising features, and ensuring alignment with business goals.

- Key skills: Business acumen, communication skills, and the ability to translate business needs into technical requirements.

Collaboration and effective communication across these roles are crucial for the success of ML Ops. The interdisciplinary nature of ML Ops teams enables organisations to tackle the complexities of deploying and maintaining ML models in real-world scenarios.

Size of ML Ops Teams

The size of an ML Ops team can vary widely based on several factors, including the scope and complexity of the organisation's initiatives, the scale of ML model deployments, and the specific requirements of the projects.

Below are some general guidelines, but keep in mind that these are just rough estimates, and the actual team size may differ depending on organisational needs.

1. *Small to medium organisation – Team size: 1 to 5 members*

Composition: This might include a mix of roles, such as a ML engineer, a data engineer, a DevOps engineer, and a software engineer. Team members may wear multiple hats in smaller organisations and handle diverse responsibilities.

2. *Medium to large organisation – Team size: 5 to 10 members*

Composition: In addition to the roles mentioned for smaller teams, there may be additional specialists, such as infrastructure specialists, QA/test engineers, security experts, and operations/support personnel. The team may be more specialised with individuals focusing on specific aspects of ML Ops.

3. *Enterprise-level organisation – Team size: 10 or more members*

Composition: Larger organisations with extensive ML deployments and a diverse range of projects may have a more extensive ML Ops team. This could include multiple ML engineers, data engineers, DevOps engineers, software engineers, infrastructure specialists, QA/test engineers, security experts, and dedicated roles for operations and support.

It is essential to note that the boundaries between roles can be fluid, and team members may collaborate closely to address the full spectrum of ML Ops tasks. Additionally, organisations may choose to leverage external ML Ops services or platforms to supplement their in-house teams, especially if they have varying workloads or specific expertise requirements.

The scale of the organisation's goal determines the size of ML operations and the complexity of the model. As ML initiatives grow, teams may evolve and scale accordingly to meet the demands of the organisation.

•

In the rapidly evolving landscape of accounting and finance, the integration of ML technologies is becoming increasingly prevalent, making it crucial for accountants to comprehend ML Ops practices. ML Ops involves deploying, monitoring, and maintaining ML models in real-world scenarios. Accountants play a pivotal role in ensuring the accuracy and reliability of financial data, and understanding ML Ops practices is essential for seamlessly incorporating ML models into financial systems. This knowledge enables accountants to contribute to the efficient deployment of automation tools for tasks like data categorisation, anomaly detection, and financial forecasting.

Moreover, familiarity with ML Ops ensures that accountants can address issues related to model performance, data integrity, and compliance with regulatory standards. As the field of accounting continues to leverage the benefits of ML, a solid understanding of ML Ops practices empowers accountants to navigate the intersection of finance and technology, contributing to more accurate financial reporting and strategic decision-making.

This chapter concludes the components of an AI ecosystem as depicted in Figure 3 which is reproduced below. Finance professionals must be proficient in understanding and implementing generally accepted and ever-evolving practices and processes in the workplace.

DATA

MODELS

INTERFACE

Ingest	Learn	Libraries
Transform	Evaluate	SDKs
Label	Deploy	APIs

ML OPS

Figure 3

10

AI Strategic Framework

An AI strategic framework is a structured approach that organisations adopt to guide their efforts in leveraging AI technologies effectively. Such a framework provides a roadmap for integrating AI into business processes, aligning AI initiatives with organisational goals, and ensuring sustainable and responsible use of AI technologies. Below is a breakdown of key components in the AI strategic framework depicted below.

1. Purpose and vision

Define the overarching purpose and vision for AI adoption within the organisation. Establish clear, measurable goals that articulate what the company aims to achieve through AI, such as enhancing operational efficiency, improving customer experiences, or driving innovation.

2. People development

Identify the skills and expertise required for successful AI implementation. Invest in training and development programs for existing employees or hire professionals with AI expertise to build a capable team.

3. Technology and data

- *Technology* – Define the AI technology stack that aligns with organisational goals. Choose appropriate frameworks, tools, and platforms for AI development and deployment. Consider factors like scalability, interoperability, and ease of integration.

- *Data strategy* – Develop a robust data strategy outlining how the organisation will collect, store, process, and manage data for AI applications. Emphasise data quality, security, and compliance with privacy regulations.

4. Processes

- *Use case prioritisation* – Prioritise AI use cases based on their potential impact on organisational objectives. Begin with projects that offer quick wins to build momentum and demonstrate the value of AI to stakeholders.

- *Assessment of readiness* – Evaluate the organisation's current technological infrastructure, data capabilities, and workforce skills. Assess the readiness of existing systems to integrate AI technologies and identify gaps that need to be addressed.

- *Implementation roadmap* – Develop a phased implementation plan for AI initiatives. Clearly outline timelines, milestones, and resource requirements for each stage. Ensure flexibility to adapt the roadmap based on feedback and evolving business needs.

5. Performance (Governance)

- *Leadership and governance* – Identify leadership roles responsible for overseeing AI initiatives. Establish governance structures to ensure ethical use, compliance with regulations, and risk management. Clearly define decision-making processes for AI projects.

- *Ethical and responsible AI* – Establish ethical guidelines for AI development and deployment. Consider the societal impact, fairness, transparency, and accountability of AI systems. Ensure compliance with legal and ethical standards throughout the AI lifecycle.

- *Collaboration and partnerships* – Foster collaboration with internal departments, external partners, and the broader AI community. Explore partnerships with AI vendors, research institutions, or industry peers to stay informed about emerging trends and best practices.

- *Performance metrics and evaluation* – Define key performance indicators (KPIs) to measure the success of AI initiatives. Regularly assess the impact of AI on organisational objectives, user experiences, and overall business outcomes. Use insights to iterate and improve AI strategies.

An effective AI strategy framework provides a holistic approach, aligning AI initiatives with the organisation's overarching goals and values. It helps organisations navigate the complex landscape of AI adoption while fostering a culture of innovation and responsible AI use. Figure 9 below depicts a framework that can be implemented in any organisation, from small to large.

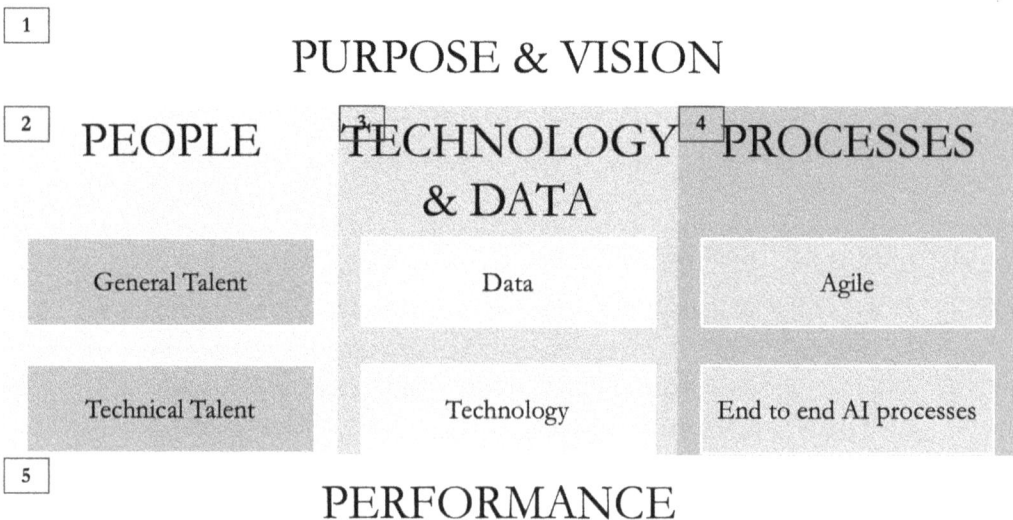

Figure 9

AI Purpose & Vision

AI purpose and vision collectively define the guiding principles and aspirations that steer the development and deployment of AI within an organisation or broader context. The AI vision encapsulates the long-term goal, outlining the desired impact of AI on the organisation's mission or industry. It serves as a beacon, inspiring stakeholders and aligning efforts toward a common trajectory. Whether it is enhancing customer experiences, revolutionising business processes, or solving complex societal challenges, the AI vision provides a roadmap for the transformative potential of AI technologies.

Complementary to the vision, the AI purpose delves into the specific objectives and values that underpin the deployment of AI. It articulates why the organisation is integrating AI, emphasising ethical considerations, responsible practices, and societal impact. The AI's purpose ensures that the technology is employed for the betterment of individuals and communities while minimising risks and fostering transparency.

A well-crafted AI vision and purpose form the foundation for responsible AI development, guiding decision-makers, developers, and users toward a future where AI aligns harmoniously with human values and collective well-being.

Examples of AI purpose and vision statements

Company: QuantumLeap Innovations

AI purpose and vision: "Empowering a quantum leap in human potential through artificial intelligence.

"We envision a world where our cutting-edge AI technologies catalyse innovation, transcend boundaries, and redefine what is possible, ushering in a future where intelligence knows no limits."

Company: Broad Solutions

AI purpose and vision: "Leading the AI revolution with ethics at its core.

"Our vision is to be a global force for good, creating artificial intelligence solutions that prioritise transparency, fairness, and societal well-being. We strive for an AI-driven future where technology aligns harmoniously with humanity."

Company: Tech Dynamics

AI purpose and vision: "Shaping a visionary future powered by intelligence.

"We aspire to be at the forefront of AI innovation, driving breakthroughs that enhance the human experience, foster sustainable practices, and inspire a world where artificial intelligence becomes an indispensable force for positive transformation."

Company: Singularity Labs

AI purpose and vision: "At the nexus of innovation and artificial intelligence, we envision a future where Singularity Labs leads the charge in unlocking the full potential of AI.

"Our vision is to create a seamless synergy between humans and machines, forging new frontiers and revolutionising industries for the betterment of society."

Company: Tech for Good Foundation

AI Purpose and Vision: "Harnessing technology for a better tomorrow.

"Our vision is to leverage artificial intelligence as a force for good, developing solutions that address pressing global challenges. We strive for a future where AI advancements contribute to sustainability, equity, and positive societal impact."

People in AI Strategy

In AI strategy, people are the linchpin that ensures the successful integration and execution of AI initiatives. Technical experts, including data scientists and ML engineers, contribute their specialised skills to design, develop, and deploy sophisticated AI models. Their expertise in algorithms, programming, and data analysis is pivotal for the success of the strategy.

Non-technical professionals, such as business analysts, project managers, and executives, also play a vital role in defining the strategic vision, identifying business needs, and aligning AI applications with organisational goals.

The collaboration between technical and non-technical teams is paramount, fostering a dynamic synergy where AI innovations are not only technically robust but also strategically aligned with the broader objectives of the company. It is the collective expertise and collaborative effort of these individuals that shape and drive the successful implementation of an effective AI strategy.

Technical staff

In crafting and executing an effective AI strategy, a harmonious collaboration between technical and non-technical staff is essential. Technical professionals play a pivotal role in developing and implementing the AI models. Their expertise in algorithms, programming languages, and data processing is fundamental to translating strategic goals into tangible AI solutions. Technical staff collaborates on tasks like data preprocessing, model training, and optimisation, ensuring the AI systems align with the organisation's objectives and industry standards.

Non-technical staff

On the other hand, non-technical staff, including business analysts, managers, and executives, are instrumental in shaping the strategic vision, defining use cases, and ensuring the ethical and responsible deployment of AI technologies. Their insights into market dynamics, customer needs, and business priorities guide the development of AI applications that generate real business value. Effective communication between technical and non-technical teams is crucial for translating complex AI concepts into actionable insights. The collaborative efforts of both technical and non-technical staff foster an environment where AI strategy is not only technologically sound but also strategically aligned with the overarching goals and values of the organisation.

AI technical staff and roles

Various technical roles contribute to different aspects of the development and deployment of AI systems. Below are some key technical roles:

1. Data scientist – They focus on extracting meaningful insights from large and complex datasets. They utilise statistical analysis, ML, and data visualisation techniques to uncover patterns and trends, informing strategic decision-making.

2. ML engineer – These engineers specialise in designing, implementing, and deploying ML models. They work on optimising algorithms, ensuring model efficiency, and integrating models into software applications.

3. AI research scientist – They are involved in cutting-edge research to advance the field of AI. They explore new algorithms, techniques, and approaches, often contributing to academic publications and pushing the boundaries of AI knowledge.

4. Computer vision engineer – These engineers focus on developing algorithms and systems that enable machines to interpret and understand visual information. Applications include image and video analysis, facial recognition, and object detection.

5. Natural Language Processing (NLP) engineer – NLP engineers work on systems that enable machines to understand, interpret, and generate ordinary language. They contribute to applications like chatbots, language translation, sentiment analysis, and text summarisation.

6. AI software developer – They build the software infrastructure for AI applications. They integrate AI

models into existing systems, develop APIs, and create user interfaces that interact with AI functionalities.

7. Robotics engineer – These engineers combine AI with robotics to design intelligent systems that interact with the physical world. They work on applications such as autonomous vehicles, drones, and industrial automation.

8. AI ethicist – They focus on the ethical considerations surrounding AI technologies. They address issues related to bias in algorithms, privacy concerns, and the responsible use of AI in various applications.

These roles often overlap, and individuals may specialise in one or more areas depending on the specific requirements of the AI project or organisation. The collaborative efforts of these technical roles are crucial for the successful development and deployment of AI solutions.

Data staff and roles

Data staff in the context of AI are professionals who specialise in managing, processing, and extracting valuable insights from data. Below are key roles related to data in AI:

1. Data engineer – These engineers focus on designing, constructing, and maintaining the architecture that allows for the processing and storage of large volumes of data. They play a critical role in ensuring the availability and reliability of data for AI applications.

2. Data analyst – These analysts examine and interpret data to provide actionable insights for decision-making. They work on identifying patterns, trends, and anomalies in the data, contributing to informed business strategies.

3. Database administrator – They manage databases that store and organise data for AI applications. They ensure

data integrity, security, and optimal performance, playing a key role in the overall data infrastructure.

4. Data architect – They design the overall structure of data systems, including databases and data warehouses. They create blueprints that guide how data is collected, stored, and utilised.

5. Data quality analyst – They focus on ensuring the accuracy, completeness, and reliability of data. They implement processes to identify and rectify data issues, maintaining high data quality standards for AI applications.

6. Big data engineer – These engineers specialise in handling and processing large volumes of data. They work with big data technologies to manage and analyse data at scale, enabling AI applications to handle massive datasets.

7. ML data engineer – This role involves preparing and structuring data specifically for ML applications, ensuring that datasets are appropriately curated and formatted to train and optimise ML models.

8. Data privacy officer – Professionals in this role focus on ensuring that AI applications handle data responsibly and in compliance with privacy regulations. They implement measures to protect sensitive information and manage risks related to data privacy.

9. Data operations manager – Data operations managers oversee the day-to-day activities related to data management. They ensure that data processes are efficient, collaborate with technical and business teams, and contribute to the overall success of AI projects.

The expertise and collaboration of data staff are fundamental for the success of AI initiatives, as high-quality, well-managed data is a cornerstone for developing and deploying effective AI models.

Non-technical staff and roles

Non-technical staff in the field of AI play integral roles in shaping strategy, aligning AI initiatives with business goals, and ensuring ethical and responsible implementation. Below are key non-technical roles in AI:

1. AI project manager – They oversee the planning, execution, and delivery of AI projects. They set project timelines and ensure that AI solutions meet business requirements.

2. AI product owner – AI product owners define the strategic vision for AI products, prioritise features based on business value, and act as liaisons between technical and non-technical teams to ensure that AI solutions align with organisational goals.

3. Business analyst – This role bridges the gap between technical teams and business stakeholders. They gather and analyse requirements, identify opportunities for AI applications, and help translate business needs into technical specifications.

4. AI strategist – AI strategists focus on aligning AI initiatives with broader business strategies. They identify opportunities for AI adoption, assess market trends, and provide insights on how AI can contribute to the organisation's competitive advantage.

5. AI governance and compliance manager – Professionals in this role ensure that AI implementations adhere to legal and ethical standards. They address concerns related to data privacy, bias, and regulatory compliance, playing a crucial role in responsible AI deployment.

6. Change management specialist – These specialists help organisations navigate the cultural and operational shifts associated with AI adoption. They focus on training, communication, and ensuring employees effectively embrace new AI technologies.

7. AI communications manager – AI communications managers are responsible for conveying the organisation's AI initiatives to both internal and external stakeholders. They play a crucial role in managing public perceptions, building trust, and highlighting the positive impact of AI.

8. AI User Experience (UX) designer – UX designers focus on creating interfaces that facilitate user interactions with AI applications. They ensure that AI solutions are user-friendly, accessible, and aligned with the expectations and needs of end-users.

9. AI education and training specialist – Professionals in this role develop training programs to enhance the AI literacy of non-technical staff. They facilitate learning sessions, ensuring that employees across the organisation understand the basics of AI and its applications.

The collaboration between technical and non-technical staff is essential for successfully integrating AI into organisations, ensuring that AI solutions not only leverage advanced technologies but also address real business needs and align with ethical and strategic considerations.

People in AI strategy are key for success, and inter-play between different roles is paramount. Most successful AI leaders and organisations that lead the AI revolution have successfully uplifted the skills of all staff on AI and related technologists.

Finance professionals play a key role in the AI ecosystem, and future professionals need to be knowledgeable and savvy to drive the respective organisations forward.

Technology and data in AI strategy

Technology and data are foundational elements in the formulation and execution of an effective AI strategy. Below is an overview of how they shape the AI strategy.

1. *Technology Stack*

- AI model development tools: Selecting appropriate tools and frameworks for AI model development is critical. The technology stack includes programming languages (e.g., Python, R), ML frameworks (e.g., TensorFlow, PyTorch), and data processing tools.

- Infrastructure and cloud services: Determining the infrastructure for AI operations is vital. Cloud services provide scalable and flexible resources, enabling organisations to efficiently deploy and manage AI applications.

- Integration with existing systems: Aligning AI technology with existing IT infrastructure is essential. Integration ensures seamless communication between AI applications and other business systems, facilitating a cohesive operational environment.

2. *Data Strategy*

- Data quality and governance: Establishing robust data governance practices ensures that data is accurate, reliable, and complies with privacy regulations. A clear data strategy outlines how data will be collected, stored, processed, and shared within the organisation.

- Data acquisition: Defining mechanisms for acquiring relevant and diverse datasets is crucial. This may involve internal data sources, external partnerships, or leveraging third-party data to enrich the organisation's dataset.

- Data security: Implementing measures to secure sensitive data is paramount. A comprehensive data strategy addresses cybersecurity concerns, ensuring that data is protected against unauthorised access and breaches.

3. Data and technology alignment

- Cross-functional collaboration: Collaboration between data and technology teams is crucial. Data scientists, engineers, and IT professionals must work closely to ensure that AI applications align with the organisation's data strategy and technical capabilities.

- Agile development practices: Adopting agile methodologies facilitates iterative development and continuous improvement. This approach allows organisations to adapt efficiently to evolving data needs and technological advancements.

- Scalability: Designing AI solutions with scalability in mind accommodates the growing demands on data and technology infrastructure. Scalable solutions can handle larger datasets and increased computational requirements as the organisation expands its AI capabilities.

4. Continuous improvement

- Monitoring and optimisation: Regular monitoring of AI models and technology infrastructure is essential. It enables organisations to identify performance issues, address potential biases, and optimise models for better accuracy and efficiency.

- Feedback loops: Establishing feedback loops between data and technology teams ensures continuous improvement. Insights gained from model performance, user feedback, and changing business needs inform updates to the data strategy and technology stack.

In summary, a well-crafted AI strategy recognises the symbiotic relationship between data and technology. An organisation's ability to collect, manage, and leverage data effectively, coupled with a technologically sound infrastructure, forms the cornerstone for successful AI implementation and long-term strategic impact.

Cloud Computing and Acceleration of AI

Cloud computing plays a pivotal role in the development, deployment, and management of AI applications.

1. Scalable infrastructure

Cloud platforms offer scalable and elastic computing resources. This is essential for handling the resource-intensive nature of AI workloads, particularly during tasks like training complex machine learning models.

2. High-Performance Computing

Cloud providers offer access to High-Performance Computing (HPC) capabilities, including specialised hardware like Graphics Processing Units (GPUs) and Tensor Processing Units (TPUs). These accelerators significantly speed up AI model training and inference.

3. Data storage and management

Cloud-based storage solutions provide a scalable and cost-effective means to store and manage large datasets. AI applications often require extensive data and cloud platforms offer various storage options to meet these requirements.

4. Serverless computing

Serverless computing models, such as AWS or Azure, simplify AI deployment. Organisations can run programming code in response to events without managing the underlying infrastructure, enhancing efficiency and cost-effectiveness.

5. AI services and APIs

Cloud providers deliver pre-built AI services and APIs, allowing organisations to easily integrate advanced AI capabilities into their applications. These services cover various domains, including natural language processing, computer vision, and speech recognition.

6. Global accessibility

Cloud services are globally distributed, enabling organisations to deploy AI applications closer to end-users. This reduces latency and improves the overall user experience, which is especially important for real-time AI applications.

7. Cost efficiency

Cloud computing operates on a pay-as-you-go model, providing cost efficiency for AI projects. Organisations pay for the resources they use, avoiding upfront infrastructure costs and allowing flexibility in resource allocation.

8. Security measures

Cloud providers implement robust security measures, including data encryption, identity management, and compliance certifications. Leveraging cloud security services helps ensure the security and privacy of sensitive AI data.

9. Development tools and environments

Cloud platforms offer a range of development tools and environments tailored for AI projects. This includes collaborative coding environments, version control systems, and Integrated Development Environments (IDEs) that streamline AI development.

10. ML platforms

Cloud providers offer dedicated ML platforms that provide end-to-end solutions for building, training, and deploying ML models. These platforms often come with tools for data preprocessing, model training, and model serving.

By leveraging cloud computing for AI, organisations gain flexibility, accessibility, and the computational power needed to innovate in the rapidly evolving field of AI without the complexities associated with managing extensive infrastructure.

Democratisation of AI

The democratisation of AI refers to the increasing accessibility and widespread adoption of AI tools, technologies, and knowledge across various industries and user groups. This trend aims to make AI capabilities more available to a broader audience beyond specialised experts or large tech companies. Below are key aspects of the democratisation of AI.

1. Accessibility

AI tools are becoming more user-friendly, with platforms and frameworks that enable individuals with varying levels of technical expertise to leverage AI functionalities. This accessibility lowers the entry barriers for businesses and professionals interested in incorporating AI into their workflows. ChatGPT is accessible to everyone which enables widespread use.

2. Cloud-based services

Cloud providers offer AI services that allow users to access pre-built models and APIs, eliminating the need for extensive AI expertise. This shift to cloud-based AI services enables organisations to integrate advanced AI capabilities into their applications without managing complex infrastructure.

3. Open-source frameworks

Open-source AI frameworks, such as TensorFlow (developed by Google) and PyTorch (developed by Meta), contribute to democratisation by providing freely available tools and resources. This allows developers and researchers worldwide to collaborate, share, and build upon each other's work.

4. Low-code and no-code platforms

The emergence of low-code and no-code AI platforms enables individuals with limited coding skills to create and deploy AI models. These platforms provide intuitive interfaces and automation to streamline the AI development process.

5. AI education and training

Educational initiatives and online courses are making AI knowledge more accessible to a broader audience. This includes training programs designed for non-technical professionals, empowering them to understand and apply AI concepts in their respective domains.

6. AI in business applications

AI is increasingly integrated into various business applications, allowing professionals in marketing, finance, human resources, and other fields to benefit from AI-driven insights and automation. This integration enhances decision-making and operational efficiency.

7. Community collaboration

Online communities and forums facilitate knowledge-sharing and collaboration among AI enthusiasts, researchers, and practitioners. This collaborative environment helps democratise AI by fostering a culture of shared learning and problem-solving.

8. Government and policy initiatives

Governments and organisations are recognising the importance of AI and implementing policies to support its responsible development and adoption. Initiatives focusing on education, research, and ethical AI practices contribute to the broader democratisation of AI.

9. Startup ecosystem

Startups play a crucial role in democratising AI by developing innovative solutions catering to specific industries or user needs. These startups often bring fresh perspectives, driving competition and diversity in the AI landscape.

10. Ethical considerations

As AI becomes more widespread, there is a growing emphasis on ethical AI practices. Ensuring fairness, transparency, and accountability in AI applications is essential to democratising AI in a responsible manner.

Overall, the democratisation of AI aims to empower a diverse range of users to harness the benefits of AI, fostering innovation and addressing societal challenges across various domains.

Processes in AI Strategy

Processes and practices are key to a successful AI strategy. Among others, agile ways of working are commonly used across organisations as a new way of organising people. It has played a key role in software development for decades. Organisations now widely use agile practices across all disciplines and divisions to drive productivity, aligned objectives, and, more importantly, employee morale and engagement.

Agile ways of working

Agile ways of working represent principles and practices that foster flexibility, collaboration, and responsiveness in project development. Whether in software development, AI projects, or other fields, agile methodologies emphasise adaptability to change and continuous improvement. Below are key aspects of agile ways of working:

1. Iterative and incremental development – Agile methodologies promote breaking down projects into small, manageable iterations. Each iteration delivers a working portion, allowing for continuous feedback and adjustments. In AI, this could involve iterative development of models, refining them based on feedback and insights.

2. Cross-functional teams – Agile teams are composed of members with diverse skills necessary to accomplish project goals. For AI projects, cross-functional teams may include data scientists, domain experts, software developers, and other relevant roles, fostering collaboration and holistic problem-solving.

3. User-centric approach – Prioritising user needs and feedback is fundamental. Agile methodologies advocate for close collaboration with end-users throughout the development process. In AI, this means understanding user requirements and adjusting models based on real-world feedback.

4. Regular feedback loops – Continuous feedback is encouraged to validate progress and make timely adjustments. Regular meetings, reviews, and retrospectives facilitate open communication. In AI development, feedback loops involve stakeholders, data scientists, and other team members to refine models and strategies.

5. Continuous integration and deployment – Agile teams focus on delivering working increments frequently. For software development, this often involves continuous integration and deployment practices. In AI, it means streamlining processes for model training, evaluation, and deployment to ensure a rapid and reliable development cycle.

6. Embracing change – Agile principles welcome changing requirements, even late in the development process. This flexibility allows teams to adapt to evolving circumstances and priorities. This adaptability is crucial for refining models and strategies in AI projects, where experimentation and learning are common.

7. Sprints and time-boxing – Projects are broken down into fixed-length iterations or sprints. This time-boxed approach provides a structured framework for planning and executing tasks. In AI, sprints may focus on specific aspects such as data preparation, model training, or evaluation.

8. Collaborative culture – Team members regularly communicate, actively share knowledge, and collaborate on problem-solving. In AI projects, this

collaborative culture is essential for bridging the gap between technical and non-technical team members.

9. Minimal Viable Product (MVP) – Agile methodologies encourage the development of a minimum viable product with essential features. This allows for early delivery, testing, and user feedback. In AI, building MVPs can involve deploying basic versions of models to gather insights and then make iterative improvements.

10. Continuous improvement – Agile methodologies emphasise a mindset of continuous improvement. Teams regularly reflect on their processes, identify areas for enhancement, and implement changes. In AI, this involves refining models, updating strategies, and incorporating lessons learned from each development cycle.

By embracing agile ways of working, teams can navigate uncertainty, respond to changing requirements, and deliver value more efficiently. The principles of agile provide a flexible and adaptive framework that can be applied across various domains, including AI projects.

Agile example in AI development

Consider the development of a customer support chatbot delivered using agile sprints.

AI Product: Customer Support Chatbot.

Sprint duration: 2 weeks

Agile Sprint 1: Define requirements (2 weeks).

Goal: Clearly define the requirements and objectives for the chatbot.

Tasks:

- Collaborate with customer support and IT teams to identify common customer queries.

- Define user stories and acceptance criteria, and prioritise features for the initial release.

- Conduct a feasibility analysis of integrating with existing customer support systems.

Agile Sprint 2: Data collection and preprocessing (2 weeks).

Goal: Collect and preprocess the data needed to train the chatbot.

Tasks:

- Gather historical customer support chat logs.

- Preprocess and clean the data to ensure quality.

- Identify and categorise frequently asked questions.

Agile Sprint 3: Model training (2 weeks).

Goal: Develop and train the chatbot model.

Tasks:

- Select and implement a suitable NLP framework.

- Train the initial version of the chatbot using the preprocessed data.

- Conduct initial testing and refinement of the model.

Agile Sprint 4: Integration and testing (2 weeks).

Goal: Integrate the chatbot into the customer support system and conduct thorough testing.

Tasks:

- Integrate the chatbot with existing customer support platforms.

- Implement APIs for seamless communication.

- Conduct end-to-end testing to ensure proper integration and functionality.

Agile Sprint 5: User interface design (2 weeks).

Goal: Design a user-friendly interface for customers and support agents.

Tasks:

- Collaborate with UI/UX designers to create an intuitive chat interface.

- Implement features like user authentication, history tracking, and multi-language support.

- Collect feedback from potential users and iterate on the design.

Agile Sprint 6: Deployment (2 weeks).

Goal: Deploy the chatbot to a controlled environment for initial usage.

Tasks:

- Set up the chatbot on a staging environment.

- Conduct a soft launch with a limited user base for initial feedback.

- Monitor performance and identify any issues that arise.

Agile Sprint 7: User feedback and iteration (2 weeks).

Goal: Collect user feedback and iterate on the chatbot's features and performance.

Tasks:

- Gather feedback from users and support agents.

- Prioritise and implement necessary improvements.

- Conduct A/B testing for new features or model enhancements.

Agile Sprint 8: Scale-Up and continuous improvement (2 weeks)

Goal: Scale up the deployment and implement continuous improvement practices.

Tasks:

- Deploy the chatbot to a wider user base.

- Implement monitoring tools for performance tracking.

- Establish a feedback loop for continuous improvement based on real-world usage.

Throughout these agile sprints, the development team collaborates closely with customer support representatives, UI/UX designers, and end-users. The iterative nature of agile methodologies allows for continuous refinement of the chatbot based on feedback and evolving requirements, ensuring the delivered AI product meets user expectations and business objectives.

Tools used in agile

Agile development relies on a variety of tools to facilitate collaboration, communication, and project management within a team. Below are some commonly used tools in agile development.

1. Jira

A widely used project management and issue-tracking tool. Jira allows teams to plan, track, and manage agile projects, including the ability to create user stories, plan sprints, and track progress.

2. Trello

A visual project management tool that uses boards, lists, and cards to organise tasks. It is particularly effective for teams that prefer a more visual and intuitive approach to project management.

3. Confluence

Often used in conjunction with Jira, Confluence is a collaboration tool for creating and sharing documents, team notes, and project documentation. It helps teams collaborate on project requirements, design documents, and other shared resources.

4. VersionOne

An agile project management tool that provides support for Scrum, Kanban, and agile methodologies. It covers various aspects of agile development, including backlog management, release planning, and reporting.

5. Asana

A versatile project management tool that supports agile workflows. Asana allows teams to create tasks, set priorities, and organise work using boards, lists, and timelines. It is suitable for both agile and non-agile project management.

6. Git and GitHub/GitLab/Bitbucket

Version control systems like Git are crucial for agile development. Platforms like GitHub, GitLab, and Bitbucket provide collaborative features for version control, branching, and merging, facilitating team collaboration on code.

7. Slack

A popular team communication tool that supports real-time messaging and collaboration. Slack helps agile teams stay connected, share updates, and maintain effective communication channels.

8. Azure DevOps

A set of development tools provided by Microsoft that includes services for version control, build automation, release management, and more. It supports agile methodologies and integrates with various development tools.

9. Miro

A collaborative online whiteboard platform that facilitates visual collaboration. Agile teams use Miro for activities like sprint planning, retrospectives, and brainstorming sessions.

10. Monday.com

A work operating system that supports agile methodologies, Monday.com allows teams to manage tasks, track progress, and collaborate on projects using customisable boards and workflows.

The choice of tools may vary depending on the specific needs and preferences of the agile team. Many teams opt for a combination of these tools to create a seamless and collaborative development environment.

There are other important processes, such as ML Ops, crucial for succeeding in an organisation's AI strategy. As ML Ops is covered in a previous chapter, the reader is encouraged to refer back to the relevant section.

Performance (Governance) in AI Strategy

Performance (Governance) in AI strategy involves establishing frameworks, policies, and processes to guide the responsible development, deployment, and management of AI within an organisation. It addresses ethical considerations, risk management, and compliance with regulations. Below are key aspects of governance in AI strategy.

1. Data & AI governance office

The data and AI governance office manages the development of policies and procedures as well as measuring the success of AI strategy. Regular meetings will cover the executive-level, managerial-level, and operational-level management of the organisation's AI strategy.

2. Policies and procedures

AI policies and procedures are essential components of an organisation's framework for responsible and ethical AI adoption. These documents articulate the principles, guidelines, and processes governing the development, deployment, and use of AI within the organisation.

AI policies typically address ethical considerations, data governance, transparency, and compliance with relevant regulations. Procedures provide a step-by-step guide for implementing these policies, outlining the practices and protocols to be followed throughout the AI lifecycle. Clear and well-defined AI policies and procedures contribute to building trust, ensuring accountability, and mitigating risks associated with AI technologies. They serve as a foundation for aligning AI initiatives with organisational values, legal requirements, and industry standards, fostering a culture of responsible AI innovation within the organisation.

Regular updates and adaptability to evolving ethical and regulatory landscapes are crucial to maintaining the relevance and effectiveness of AI policies and procedures.

3. Measurement

The measurement of an AI strategy involves a comprehensive evaluation of its impact on organisational objectives, efficiency gains, and the overall alignment with business goals. A crucial metric is the return on investment (ROI), assessing the financial outcomes relative to the costs incurred in implementing AI initiatives. This includes quantifying the tangible benefits such as increased revenue, cost savings, or improved operational efficiency derived from AI applications.

Additionally, assessing the performance of individual AI models and applications is essential, examining factors like accuracy, reliability, and user satisfaction to ensure they at least meet expectations. By aligning the measurement process with business objectives and tracking key performance indicators, organisations gain valuable insights into the strategic value of their AI investments.

Furthermore, the measurement of an AI strategy extends beyond financial metrics to encompass ethical considerations, regulatory compliance, and the organisation's ability to adapt to evolving technological landscapes. Evaluating the strategy's agility involves assessing its responsiveness to changing business environments and emerging opportunities.

Ethical and regulatory compliance metrics ensure that AI applications adhere to established guidelines, safeguarding data privacy and maintaining public trust. The effectiveness of risk management processes associated with AI projects is another critical dimension, ensuring that potential risks are identified, assessed, and mitigated appropriately. Continuous improvement and learning mechanisms, guided by feedback and lessons learned from AI implementations, contribute to the refinement and optimisation of the overall AI strategy over time.

11

Case Study 1 – Amazing Banking Corporation & Digital Banking

Names, locations, and specific financial data have been changed to protect privacy.

Background

Amazing Banking Corporation (ABC) began as a bank based in New York in the early 1900s and has played an important role in the history of US banking from the Civil War to the Great Depression and the Financial Crisis of 2008. In the 2000s, ABC became a true finnancial behemoth, with many banking operations such as retail banking, investment banking, insurance, and asset management. ABC's mission was to offer everything financial to everyone worldwide.

ABC has a large customer base of a few hundred million globally and a workforce of more than 300,000. In 2020, ABC was struggling to maintain the profitability of the business and continuously reported lower than the guidance to the market.

Challenge

ABC recognised that it had yet to be modernised to face the realities of the 21st century. Its IT infrastructure was old, mostly with mainframe computers that were not suitable for the digital age. ABC also recognised the need to modernise the ways of working with its global workforce of 300,000. ABC estimated that old IT infrastructure added a 20% extra cost burden compared to its peers with a modern technology stack.

ABC's retail banking unit relied heavily on the branch network and staff without a comprehensive modern digital ecosystem.

Investment banking and asset management businesses operated through accredited fund managers and brokers who added extra complexity and cost to the business without a proper digital ecosystem.

At its core, ABC was a large behemoth from the industrial era that struggled to survive in the digital and AI age.

Rising above the Challenge

In response to the dynamic shifts within the financial industry and the evolving expectations of modern consumers, ABC undertook a visionary transformation. Recognising the imperative to adapt to a rapidly changing landscape, the company strategically embraced AI as the catalyst for its metamorphosis into a global industry leader. The decision was fuelled by an acute awareness of the profound impact AI could wield in navigating complexities, optimising processes, and elevating customer experiences to a new level.

In the face of intensifying competition and a digitally driven era, ABC embarked on a journey marked by innovation and technological prowess. The infusion of AI technologies into its core operations became a cornerstone of this transformation, promising not just efficiency gains but a fundamental change in the financial services it offered. The company's leadership understood that harnessing the power of AI wasn't merely a response to contemporary challenges but a strategic imperative to proactively shape the future of finance.

The strategic integration of AI not only enhanced operational efficiencies but also positioned the company at the forefront of pioneering customer-centric financial services.

By aligning its operations with the potential of AI, the company not only addressed current challenges but anticipated future trends, solidifying its position as a frontrunner in the finance industry. This transformative journey serves as a testament to the company's resilience, adaptability, and strategic foresight in navigating the complexities of the modern financial ecosystem.

Strategy and the Scorecard

Embarking on a strategic review, ABC conducted a thorough examination of its current position, market dynamics, and future opportunities. Recognising the need for a roadmap that aligns with its vision, the bank meticulously developed a comprehensive strategy underpinned by a strategic scorecard. This scorecard served as a dynamic tool to measure and track KPIs, ensuring a transparent and accountable approach to the transformative journey ahead.

The strategic review facilitated a deeper understanding of the bank's strengths, weaknesses, opportunities, and threats. The subsequent development of the strategy involved a collaborative effort, engaging key stakeholders and drawing insights from diverse perspectives. The strategic scorecard, a visual representation of the bank's objectives and corresponding metrics, provided a roadmap for success.

Aligned with the vision, the scorecard encapsulated digital transformation, customer-centric innovations, risk management, and sustainable growth. It served as a living document that not only guided decision-making but also facilitated regular assessments and adjustments, ensuring agility in response to market dynamics.

Through this strategic review and the implementation of a dynamic scorecard, ABC was able to navigate the complexities of the financial landscape with precision. The bank's commitment to transparency, innovation, and strategic accountability positioned it for sustained success as it journeyed into a future marked by financial empowerment and unparalleled customer experiences.

Figure 10 below is ABC's strategic scorecard.

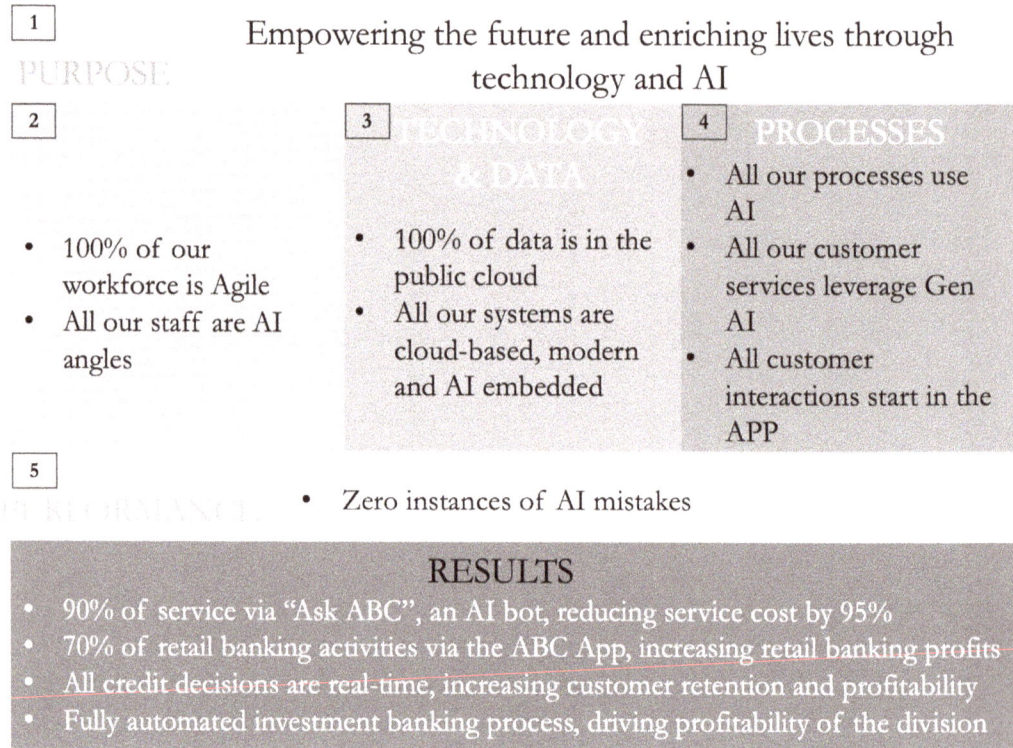

Figure 10

How ABC Leveraged AI to Transform Its Business

Generative AI to transform customer service and knowledge management

Leveraging the transformative power of generative AI, ABC has pioneered the development of a cutting-edge customer service tool named "Ask ABC". This innovative tool was designed to revolutionise the way customers interact with the bank by providing swift and accurate responses to their queries. ABC leveraged one of the off-the-shelf Generative AI tools available in the market to fast-track its implementation and keep pace with the innovation in the area.

"Ask ABC" harnessed the capabilities of generative AI to comprehend natural language and generate contextually relevant answers. By employing advanced language models, the tool ensured a seamless and intuitive conversational experience for customers seeking information about their accounts, transactions, or banking services. This generative AI-driven tool not only enhanced efficiency but also reflected ABC's commitment to leveraging emerging technologies for the benefit of its customers.

By integrating generative AI, the bank was not only enhancing the accessibility of its customer service but also setting new standards for responsiveness and user engagement in the banking industry. This transformative tool underscored the bank's dedication to staying at the forefront of technological advancements, providing customers with an intelligent and efficient avenue for obtaining information and support.

Open data, AI and real-time integration make the customer experience seamless

The bank redefined the customer experience by seamlessly integrating open banking and AI to offer real-time credit decisions. This innovative approach not only enhanced the speed and accuracy of credit assessments but also created a customer-centric banking experience that aligns with the modern, interconnected financial landscape.

Through open banking, ABC tapped into a wealth of customer-consented financial data from various sources, including third-party financial institutions and service providers. This comprehensive data ecosystem provided a holistic view of a customer's financial health, enabling the bank to make informed credit decisions.

The integration of AI algorithms into this open banking framework ensured that the credit assessment process is not only data-driven but also adaptive and responsive. ML models analysed diverse sets of financial data in real-time, considering not only traditional credit scores but also transactional patterns, spending behaviours, and other relevant factors. This nuanced approach allowed the bank to tailor credit decisions to individual customer profiles, providing a more accurate representation of creditworthiness.

The real-time aspect of this process was a game-changer for the customer experience. Customers applying for credit can receive decisions almost instantly, eliminating prolonged waiting times and enhancing the overall convenience of banking interactions. This speed was made possible by AI-driven automation, which rapidly processed and analysed the vast amount of data available through open banking.

This has significantly reduced ABC's cost of doing business, thus improving the overall profitability of its core functions. Prior to the implementation of this solution, the average turnaround time of credit files was 5 business days for a simple credit decision.

Computer vision technologies and staff-less branch

Today, ABC has a network of branches that operate as staff-less branches. These branches represent 50% of the branch network and 75% of retail transaction volume.

The bank has revolutionised traditional banking by harnessing the power of computer vision and AI to establish a groundbreaking staff-less branch network. This innovative approach redefines the banking experience, combining cutting-edge technology with seamless customer interactions.

The incorporation of computer vision technology enables the staff-less branches to operate efficiently and securely. Advanced cameras and sensors are strategically placed within the branch environment, allowing for real-time monitoring of customer movements and activities. These computer vision systems are integrated with AI algorithms that can identify and authenticate customers, ensuring a secure and personalised experience.

AI plays a pivotal role in automating routine banking tasks within staff-less branches. Customers can perform various transactions, such as cash withdrawals, deposits, and account inquiries, through intuitive AI-powered interfaces. Natural language processing capabilities allow customers to communicate with the AI systems and to conversationally receive assistance and information.

In addition to transactional capabilities, the AI-driven system is equipped to provide personalised financial advice and product recommendations based on individual customer profiles and preferences. This creates a tailored experience comparable to that of interacting with a human banker.

The staff-less branch network enhances operational efficiency. Routine tasks are automated, reducing wait times for customers and optimising resource allocation. The AI algorithms continuously learn and adapt to customer behaviours, ensuring that the banking experience evolves in response to changing needs.

Security and privacy are paramount in this staff-less banking model. The combination of computer vision for secure entry and AI for transactional processes is designed to meet the highest standards of data protection and customer confidentiality.

By leveraging computer vision and AI to create a staff-less branch network, ABC not only embraces technological innovation but also redefines the way customers engage with their financial institutions. This forward-thinking approach aligns with the bank's commitment to providing efficient, secure, and customer-centric banking services in an increasingly digital era.

Investment banking and robo-advisor

ABC has strategically integrated robo-advisors into its investment banking services, ushering in a new era of efficiency, accessibility, and personalised wealth management. This innovative approach leverages AI to provide clients with sophisticated investment advice, portfolio management, and financial planning, all tailored to individual goals and risk profiles.

Robo-advisors at ABC are powered by advanced ML algorithms that analyse vast datasets, market trends, and economic indicators in real-time. These algorithms not only assist clients in making informed investment decisions but also adapt to changing market conditions, ensuring a dynamic and responsive investment strategy.

Clients interact with the robo-advisors conversationally through intuitive digital interfaces that utilise natural language processing. This enables a user-friendly experience where clients can articulate their financial goals and preferences and receive personalised investment recommendations promptly. The robo-advisors consider factors such as risk tolerance, investment horizon, and market conditions to propose optimal asset allocations and investment strategies.

The three-tiered approach – small, medium, and large – requires different levels of approvals. Small trades that are below a certain threshold based on the client's experience are executed automatically. Medium trades require confirmation by the client. Large trades that are above a certain threshold will be executed upon approval from the client and a human advisor.

Automation is a cornerstone of the robo-advisor model, allowing for the swift execution of investment decisions without manual intervention within the parameters described above. This not only reduces operational costs but also ensures that clients can capitalise on market opportunities in a timely manner.

More importantly, the integration of robo-advisors aligns with ABC's commitment to financial inclusivity. The platform caters to a broader spectrum of investors, including those with smaller portfolios, providing them access to sophisticated investment strategies that were traditionally reserved for high-net-worth individuals.

Core platform upgrade and data modernisation

ABC has taken a significant leap forward in its technological evolution with the core platform upgrade to a cloud-based architecture, accompanied by the integration of modern data and AI tools. This strategic move positions the bank at the forefront of the digital banking landscape, enhancing operational efficiency, scalability, and agility. Migrating to a cloud-based infrastructure ensures seamless access to computing resources, enabling the bank to scale its operations dynamically based on demand while optimising costs.

This core platform upgrade not only represents a technological milestone for the organisation but also underscores its commitment to innovation and customer-centricity. The tools facilitate the processing and analysis of vast datasets, empowering the bank with actionable insights into customer behaviour, market trends, and operational performance. Additionally, these tools enhance decision-making processes, offering personalised financial services, real-time analytics, and predictive capabilities, ultimately delivering a superior and more responsive banking experience for customers.

By embracing cloud-based architecture and cutting-edge AI and data tools, the bank positions itself to adapt swiftly to industry shifts, comply with regulatory requirements, and continuously refine its services based on evolving customer needs.

People were at the core of the transformation

This transition involved cross-functional teams working collaboratively in short, focused sprints, fostering continuous feedback loops and rapid decision-making. By embracing agile ways of working, ABC not only enhanced its ability to deliver projects efficiently but also fostered a culture of innovation and continuous improvement.

The bank's strategic transformation gained momentum as the agile framework facilitated a more dynamic and customer-centric approach, ensuring that the organisation remained agile, responsive, and well-positioned for sustained success in the ever-evolving banking landscape.

Moreover, ABC rolled out an employee accreditation program that covered basic, advanced, and expert-level curricula on AI. This was a mandatory program that required each employee to complete the level of accreditation required by their job. In order to achieve traction and adoption, ABC leadership incorporated accreditation as a minimum requirement for annual staff incentives.

Impact on Finance

The CFO of the bank initiated a finance transformation program that modernised its financial systems, resulting in significant advancements and efficiencies in several key areas:

1. Real-time processing of financial transactions – This capability shift not only improved the accuracy of financial records but also enhanced the bank's responsiveness to market changes and customer needs.

2. Annual planning and budgeting – By leveraging advanced data analytics and forecasting models for annual planning and budgeting, the bank was able to predict future financial trends more accurately and allocate resources more efficiently.

3. Risk management – The upgraded systems provided more sophisticated tools for risk assessment and management. The bank could now instantly analyse and respond to various financial risks, such as credit, market, and operational risks, thereby protecting its assets and customer investments more effectively.

4. Month-end closing – The modernised finance system significantly reduced the lead time for month-end closing activities. Automated data aggregation and processing capabilities enabled faster reconciliation, reporting, and analysis, leading to timely financial statements and insights.

5. Internal auditing – Automated auditing tools powered by AI and ML could quickly identify discrepancies and potential areas of concern, ensuring higher compliance standards and financial integrity.

6. Taxation – The new system could automatically calculate and process tax liabilities, ensuring compliance with tax laws and regulations while minimising errors and reducing the workload for the finance team.

Overall, the finance transformation program at ABC represented a significant step forward in the bank's operational capabilities. By modernising its systems, the bank not only achieved real-time processing and reduced lead times in key financial processes but also enhanced its overall strategic decision-making and competitiveness in the financial sector.

Final Outcome

Two and a half years into the three-year strategy, ABC announced that it had achieved all of its objectives successfully. Figure 11 below is the scorecard ABC released to the market and its shareholders.

#	Objective	Result	Outcome
1	Double digit revenue growth	12%	Exceeded
2	Single digit EBITDA growth	8%	Exceeded
3	Maintain dividends	Maintained	Met target
4	Fully Agile workforce	100%	Met target
5	Fully AI accredited workforce	100%	Met target
6	Cloud-based IT infrastructure	100%	Met target
7	100% AI in business processes	100%	Met target
8	Data in public cloud	95%	Below target
9	Ask ABC service level across the business	92%	Exceeded
10	X million customers on the ABC App	X.2m	Met target
11	Staff-less branch network	51%	Met target
12	Automated Investment Banking	100%	Met target
13	Brand recognition score	Top 100	Exceeded
14	Employee Experience Score	Top tire	Exceeded
15	Customer Experience Score (NPS)	45	Exceeded

Figure 11

12

Case Study 2 – Tiny Inc. Transforms Its FP&A

Names, locations, and specific financial data have been changed to protect privacy.

Background

Tiny Inc., a global conglomerate headquartered in London, exemplifies a multifaceted and expansive business model. Its operations, sprawled across all continents, encompass a diverse range of industries, making it a prominent player in the global market.

Diverse business sectors – Tiny Inc. has established a strong presence in seven sectors, covering telecommunications, transport, finance, food and beverage, retail, mining and leisure. This diversification not only provides a broad revenue base but also mitigates risks associated with market fluctuations in individual sectors.

Global operations – The company's reach is truly global, with subsidiaries and associate companies located around the world. This international footprint allows Tiny Inc. to tap into local markets, adapt to regional trends, and cater to various customer needs.

Large workforce – Distributed across various service locations, it contributes to the local economies while ensuring that the conglomerate's operations run smoothly and efficiently.

Headquarters in London – Based in a major global financial hub, Tiny Inc. maintains strategic advantages in terms of access to capital markets, global connectivity, and a favourable business environment.

Cultural diversity and inclusion – With a workforce and operations in multiple countries, Tiny Inc. is positioned to embrace cultural diversity, bringing together many perspectives and skills that can foster innovation and creative problem-solving.

Economic impact and corporate responsibility – As a conglomerate, Tiny Inc. likely plays a significant role in the economies of the countries it operates in. This position comes with a responsibility to engage in sustainable practices and contribute positively to the communities it serves.

In summary, Tiny Inc.'s expansive and diversified operations, coupled with its strategic headquarters in London and its significant workforce, position it as a key player in the global business landscape, with a presence that spans across continents and industries.

Challenge

The Financial Planning & Analysis (FP&A) function at Tiny Inc. faced several challenges due to the current structure and tools used in its finance planning processes. These challenges, stemming from reliance on traditional methods and tools, significantly impacted the efficiency and accuracy of their financial planning.

1. Extended duration of annual planning process – The annual planning process, starting in January and spanning over six months, reflected the complexity of Tiny Inc.'s operations. However, this extended duration resulted in outdated plans by the time they were completed, reducing their relevance and effectiveness.

2. Dependence on company finance functions – The FP&A team relied on quarterly plans from over 100 companies with diverse operations so coordinating and consolidating this information in a timely manner was complex.

3. Over-reliance on the planning team – The FP&A team's dependence on the planning team for rolling forecasts added another layer of complexity. This reliance created bottlenecks, especially if the planning team was overwhelmed or lacked resources.

4. Use of MS Excel for coordination – While Excel is a versatile tool, its use for coordinating plans across over 100 companies was not ideal. Excel's limitations led to inefficiencies, planning errors, and difficulties in managing complex datasets.

5. Employee dissatisfaction – The long and tedious finance planning process, characterised by manual handling and a lack of modern tools, led to employee dissatisfaction. This dissatisfaction stemmed from the repetitive nature of the tasks, the likelihood of error, and the feeling of being unequipped to handle such a large-scale operation effectively.

6. Planning errors and inaccuracies – Manual data handling, especially in a complex environment like Tiny Inc., increased the likelihood of errors. These errors propagated through the planning process.

7. Lack of modern planning tools – The absence of modern, integrated financial planning tools limited the FP&A team's ability to process, analyse, and report financial data efficiently.

The Head of Planning of Tiny Inc. estimated that more than 200,000 staff hours were spent on the planning process, equating to millions of dollars every cycle.

To address these challenges, the CFO of Tiny Inc. undertook a transformation in the FP&A function with the support of the global leadership team. The CFO set the ambitious objective of having always-ready plans across its operation to reduce the financial planning process from six months to two weeks.

FP&A Transformation

The finance transformation team at Tiny Inc. embarked on a comprehensive analysis of the company's ecosystem to propose a future-state architecture. This new architecture was designed to be a holistic solution encompassing data management, AI models, and seamless integration into existing business processes. Below is an overview of the key components of the proposed future-state architecture.

1. Data management infrastructure

- Centralised data repository – Establish a unified data repository to consolidate data from various sources of the group of companies, ensuring data consistency and accessibility.

- Data quality and governance – Implement robust data quality management and governance frameworks to ensure accuracy, security, and compliance with regulatory standards.

- Advanced analytics capabilities – Integrate tools for advanced data analytics to derive actionable insights and support data-driven decision-making.

2. AI models

- Predictive analytics and ML – Develop AI models that can predict financial trends, customer behaviour, and market dynamics, enhancing forecasting accuracy.

- Automated decision-making tools – Utilise AI to automate routine decision-making processes, thereby increasing efficiency and reducing the scope for human error.

- Custom AI solutions – Create tailored AI solutions for specific finance functions such as revenue recognition, product-price estimates, tax calculations, and expense forecasting.

3. Integration into business processes

- Seamless integration with existing systems – Ensure the new architecture integrates smoothly with existing IT infrastructure and business applications.

- User-friendly interfaces and dashboards – Design intuitive user interfaces and dashboards that allow easy access and manipulation of data and AI models by finance professionals.

- Training and change management – Implement comprehensive training programs and change management strategies to facilitate the adoption of new technologies and processes by employees.

4. Scalability and flexibility

- Modular design – Adopt modular design that allows for scalability and flexibility, enabling the architecture to evolve with changing business needs and technological advancements.

- Future-proofing – Considering future trends in AI and data analytics to ensure the architecture remains relevant and effective in the long term.

5. Regulatory compliance and ethical considerations

- Compliance with financial regulations – Ensure all aspects of the architecture comply with international and local financial regulations.

- Ethical AI use – Incorporate ethical considerations in the development and deployment of AI models, particularly regarding data privacy and bias mitigation.

By implementing this future-state architecture, Tiny Inc. aimed to revolutionise its finance function, making it more data-centric, efficient, and responsive to the dynamic demands of the global financial environment. Figure 12 below is the new landscape of the FP&A ecosystem.

AI transformation of FP&A landscape

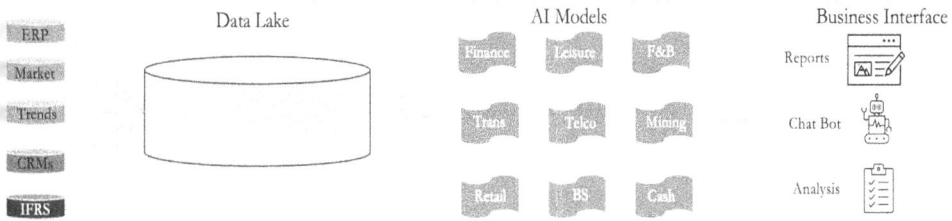

Figure 12

6. Thousands of Excel data sheets to a data lake

The implementation of a comprehensive data strategy by Tiny Inc., including the establishment of an enterprise data lake, marked a significant shift in the company's approach to data management and utilisation. This strategic move was aimed at overcoming the limitations posed by the use of manual data sheets and an over-reliance on spreadsheet files.

- Enterprise data lake implementation – The data lake was designed to consolidate all relevant data, both financial and non-financial, from across the subsidiaries. This centralised repository enabled more efficient data storage, processing, and retrieval, facilitating better data analytics and decision-making.

- Overcoming initial reluctance – Finance staff, accustomed to traditional tools like Excel, initially showed reluctance to transition to the new system. This hesitation was rooted in familiarity and concerns over adapting to a new technology, which often involves a steep learning curve.

- Trial with select companies – To demonstrate the effectiveness of the new system, a trial was conducted with a few companies for staff to witness firsthand the advantages of the data lake over traditional Excel sheets.

- The new system proved to be more reliable, reducing errors associated with manual data handling and spreadsheet management. Improved data processing speed and the ability to handle large datasets efficiently were key benefits. The system allowed for more sophisticated data analysis techniques, such as predictive analytics and real-time reporting. The data lake provided scalability, enabling Tiny Inc. to easily incorporate data from new sources or subsidiaries as the company grows.

- Change management and training – Tiny Inc. invested in change management and training programs to ease the transition. These programs focused on highlighting the benefits of the new system and providing the necessary skills and support to the finance staff.

- Tiny Inc.'s strategic move to replace manual and Excel-based data management with an enterprise data lake initially faced resistance. However, the successful trial and the tangible benefits observed convinced the finance staff of its effectiveness, leading to a more capable, reliable, efficient, and scalable data management system that significantly improved the overall operation of the business.

7. Automated forecasting driven by AI

The transformation of the financial planning process at Tiny Inc. was a significant undertaking that involved a shift from a labour-intensive approach to a more sophisticated, AI-driven methodology. This transition addressed the challenges posed by the involvement of a large number of finance staff, who had varying levels of experience, knowledge, and technical skills.

- Challenges with the traditional planning process – Variabilities in staff expertise and knowledge often led to inconsistencies in the financial plans. The process was time-consuming and often resulted in plans that lacked robustness and accuracy.

- Need for industry-specific AI models – Tiny Inc. recognised the necessity for industry-specific AI models to improve the accuracy and reliability of its forecasting for the entire group sectors.

- Formation of a specialised data science team – The company assembled a team of data scientists specifically chosen for their industry knowledge and expertise in AI and data analytics. This team's primary goal was to develop, test, and deploy AI models tailored to the company's forecasting needs.

- Development and deployment – The data science team worked tirelessly to create models that could accurately predict financial outcomes based on a myriad of variables and historical data. Rigorous testing was conducted to ensure the reliability and accuracy of these models before they were deployed across the group.

- Outcomes of the AI-driven approach – The new AI models consistently provided more accurate and reliable forecasting outcomes with a much shorter turnaround time. This shift not only enhanced the efficiency of the planning process but also reduced the workload on the finance staff, allowing them to focus on more strategic aspects of financial planning.

The introduction of industry-specific AI models for financial forecasting at Tiny Inc. marked a pivotal step in the company's journey towards modernising its financial planning processes. This approach not only improved the accuracy and efficiency of financial forecasts but also streamlined the overall planning procedure, aligning it with the company's strategic objectives and future growth plans.

8. No more consolidation of Excel files as the system generates summaries at all levels

The transition to the new system at Tiny Inc. marked a significant improvement in the reporting process within the corporate finance team. This change addressed the key issues that were prevalent with the previous manual system, enhancing both the efficiency and quality of the reports.

- Challenges with the previous manual system – The corporate finance team previously relied on consolidating data from various manual files to generate management reports. This process was not only time-consuming but also prone to manual errors, leading to inaccuracies in the reports. The delay in producing reports due to the manual consolidation process often resulted in management receiving outdated or delayed information.

- Efficiency of the new system – The new system automated the data consolidation process, significantly reducing the time required to generate reports. Automation minimised the occurrence of manual errors, ensuring more accurate and reliable reporting. The ability to produce reports more quickly meant that management could receive timely information, which is crucial for strategic decision-making.

- Improved report content – The new system was capable of preparing summaries that catered to all levels of management, providing both high-level overviews and detailed analyses as needed. It included supporting documents such as market trend summaries and inflation trends, offering a more comprehensive view of the financial landscape. The system also incorporated country-specific assumptions, which were particularly important for a global conglomerate like Tiny Inc., ensuring that reports were relevant and tailored to specific regional contexts.

- Insightful financial planning documents – The reports generated by the new system provided a deeper understanding of various financial metrics and their implications on the company's performance and strategy. This level of detail and insight supported more informed decision-making at all management levels.

The implementation of the new system in the corporate finance team at Tiny Inc. transformed the management reporting process. By automating data consolidation and report generation, the system not only reduced manual errors and delays but also enhanced the quality and depth of management reports. This improvement played a crucial role in supporting more strategic and informed decision-making across all levels of management in the company.

Final Outcome

The successful transformation of the FP&A function at Tiny Inc. led to a remarkable improvement in the efficiency and effectiveness of its financial planning process.

1. Dramatic reduction in the planning cycle – The most significant achievement was the reduction of the planning cycle from a lengthy six months to a swift two weeks. This was accomplished through the implementation of advanced data management systems, AI-driven forecasting models, and streamlined processes as outlined above.

2. Boosted morale – The shortened planning cycle and the adoption of more efficient tools and practices greatly improved the morale of the finance staff. Freed from the burdens of a protracted and manual-intensive process, they could focus on more strategic, value-added activities.

3. Improved accuracy – The transformation led to a marked improvement in the accuracy of financial forecasts and plans. The use of sophisticated data analytics and AI models allowed for more precise and reliable financial projections.

4. Enabling timely critical decisions – The ability to produce financial plans and forecasts quickly enabled management to make critical decisions promptly. This agility in decision-making is crucial in the fast-paced global business environment, where opportunities and risks can arise swiftly.

5. Competitive advantage – As reported by the CFO of Tiny Inc., this transformation has provided the company with a significant competitive edge. The enhanced speed, accuracy, and efficiency in financial planning have positioned Tiny Inc. to respond more effectively to market changes and to capitalise on new opportunities.

6. Overall impact on the business – The FP&A transformation has had a ripple effect across the organisation, impacting various aspects of the business positively. It has led to better resource allocation, improved financial performance, and a stronger alignment of financial strategies with the company's overall business goals.

In conclusion, the transformation of the FP&A function at Tiny Inc. has been a game changer for the organisation. It not only streamlined the financial planning process but also empowered the company to operate more dynamically and strategically in the global market. This improvement in internal operations and decision-making capabilities has positioned Tiny Inc. as a more agile and competitive player in the global market.

Figure 13 below is the strategic scorecard of the FP&A transformation program as published by Tiny Inc.

1 PURPOSE

Empowering financial decisions at the speed of light, in every business endeavour

2

- All finance staff are proficient in AI technologies

3 TECHNOLOGIES & DATA

- All data in the cloud-based central repository
- Segment/country-specific AI algorithms empowering planning

4 PROCESSES

- Financial planning cycle reduced to 2 weeks
- Real-time forecasts accessible to business leaders

5

- 100% data quality in financial data

RESULTS

- 100,000 hours of planning time saved across the group
- Insight to decision time reduced to 5 days
- Overall FP&A cost reduced by 80%

Figure 13

13

Case Study 3 – Tootal Telekom & Finance Processes

Names, locations, and specific financial data have been changed to protect privacy.

Background

Tootal Telekom, a large telecommunication company headquartered in a Western European city, is a classic case of a legacy organisation that transitioned to meet the demands of the digital age. The company's journey from being a government-owned entity to a modern digital enterprise involves several key challenges and opportunities.

Inherited from its time as a government-owned entity, Tootal Telekom had established processes and systems that were not fully aligned with the rapid pace and innovation required in today's digital landscape. These systems were less flexible or scalable compared to newer technologies.

To remain competitive and relevant, the company embarked on a comprehensive digital transformation journey. This involved not just upgrading technology but also rethinking business processes, customer interactions, and internal operations to be more agile and customer-centric.

Challenge

The finance department of Tootal Telekom was entrenched in a cycle of outdated methodologies, heavily relying on manual procedures and legacy systems to compile their monthly financial statements. This approach was heavily dependent on time-consuming manual data entry and legacy systems that lacked the agility and efficiency of modern financial software.

The critical juncture in their monthly financial cycle is Work Day 10, a designated date in the monthly financial calendar. It often falls into the third week of the following month, when the draft financial statements are prepared for executive review. However, this system's inherent delays often led to surprises for the senior management and leadership, as the financial results frequently showed significant deviations from the established budgets and forecasts.

This process's inefficiency was further highlighted by the substantial time lag in responding to these financial discrepancies. By the time the finance team conducted a thorough analysis and understood the underlying causes of these variances, almost an entire month had elapsed. Consequently, the timeframe to devise and implement effective corrective measures stretched to nearly three months. During this period, the company remained unable to take timely tactical actions to address the financial issues, leading to prolonged periods of operational inefficiency and potentially missed opportunities for strategic financial adjustments.

This sluggish response rate underscored the critical need for Tootal Telekom to overhaul its financial processing methods, transitioning towards more agile, technology-driven systems that could provide real-time insights and faster decision-making capabilities.

Project Titan

Tootal Telekom's CEO recognised the need for the finance organisation to evolve to align with the company's strategic vision. This led to a decision to embark on a comprehensive transformation of the financial close process, an initiative aimed at accelerating the process while simultaneously enhancing user experience. To spearhead this ambitious project, named "Project Titan", the finance team was brought together to work collaboratively, aiming to establish a best-in-class finance close process. This project was not just about improving efficiency but also about redefining the finance function's role within the company.

The scope of Project Titan was broad and meticulously planned. It included setting clear performance targets, which were essential to measure the project's success. This ensured that the team's performance was consistently aligned with the set targets. The transformation project was structured to focus on several key areas:

- Zero-day close involved refining local closing procedures to ensure that financial records were ready before the end of the month for the close process.

- Calendar and governance aimed at optimising the financial closing schedule and ensuring compliance.

- Finance systems, consolidation, and governance, which focused on upgrading financial systems for better data consolidation and control.

- External reporting to enhance the accuracy and timeliness of financial information shared outside the company.

Each of these areas was critical to achieving a holistic transformation of the finance function, moving it away from outdated processes and towards a more dynamic, efficient, and responsive model. This shift was crucial for Tootal Telekom to not only keep pace with the rapidly changing business environment but also to position itself as a leader in financial management efficiency and transparency.

Real-time Data Ingestion

A key objective for Tootal Telekom was to achieve a zero-day close, a goal requiring significant enhancements in data processing. The team pinpointed near real-time data ingestion as an essential element in revamping the close process. The existing system architecture at Tootal Telekom, primarily designed for monthly processing of various systems – such as billing, payment, and collection – was identified as a significant bottleneck in achieving this objective. The legacy infrastructure was not capable of supporting the desired rapid data processing and analysis.

To address this challenge, the team embarked on deploying a new, more agile API-based architecture that provided seamless connectivity across all systems, linking them to the company's recently upgraded financial system. This was to ensure faster and more efficient data transfer and processing.

A critical component of this transformation was the development of a data lake. This was designed to collect and ingest data from various sources within the organisation, offering a centralised repository for all financial data. The data lake's architecture was tailored to not only address current needs but also to be future-proof, accommodating potential changes and expansions in data requirements. This foresight was crucial in ensuring that the data infrastructure would remain relevant and effective in the face of evolving business demands and technological advancements.

In short, Project Titan's strategy for the company was a comprehensive overhaul of its data processing capabilities, moving from a periodic legacy system to a dynamic, API-driven architecture complemented by a robust data lake. This strategic shift was pivotal in realising the ambitious goal of a zero-day close, setting a new standard for efficiency and responsiveness in the company's financial operations.

Figure 14 below illustrates the newly revamped process flow that transformed Tootal Telekom's close process.

Revamped Zero-Day Close

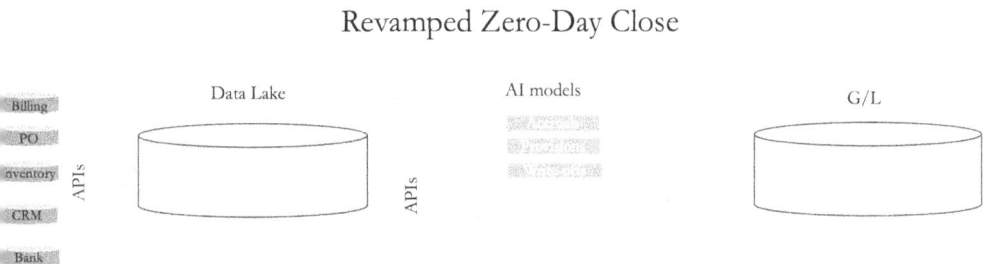

Figure 14

Automated Journals and Calculations

Tootal Telekom's initiative to enhance the speed and accuracy of its financial closing process involved the strategic implementation of AI algorithms. These algorithms were specifically tailored to manage and automate complex financial tasks such as accruals, prepayments, write-offs, and other critical financial adjustments.

The adoption of AI algorithms brought two primary benefits to Tootal Telekom's finance function. First, it significantly accelerated the financial closing process. By automating tasks that traditionally required extensive manual effort and time, AI algorithms enabled faster compilation, processing, and reconciliation of financial data. This acceleration was crucial in achieving more timely financial reporting, a key competitive advantage in today's fast-paced business environment.

Second, the use of AI also enhanced the accuracy of financial statements. AI algorithms are adept at handling large volumes of data with precision, reducing human error typically associated with manual processes. These algorithms can also identify patterns and anomalies that might be overlooked in manual reviews, ensuring more accurate financial reporting. This increased accuracy is vital for reliable financial planning and decision-making, as well as for maintaining compliance with regulatory standards.

Overall, Tootal Telekom's integration of AI into its financial processes represented a forward-thinking approach, leveraging cutting-edge technology to optimise its financial operations. The success of this initiative demonstrates the potential of AI to transform finance functions, offering a blueprint for other organisations seeking to modernise their financial processes.

End-to-end Visibility

Tootal Telekom's decision to implement a world-class visualisation tool for all staff significantly transformed the way information is accessed and analysed within the company. This strategic move not only streamlined the report preparation process but also empowered employees across various departments to engage more deeply with data.

Visualisation tools are known for their ability to present complex datasets in a more accessible format, such as graphs, charts, and interactive dashboards. This approach allowed staff to quickly grasp key trends, patterns, and insights without the need for extensive data processing skills.

One of the most notable impacts of this tool was the reduction in report preparation time. By providing a user-friendly interface for data analysis, finance staff could shift their focus from the time-consuming task of preparing reports to value-adding activities like business analysis and decision support.

Furthermore, the tool's drill-down capabilities offered staff the ability to independently explore data at their convenience. This feature was particularly beneficial as it allowed for a deeper analysis of specific areas of interest. Staff could compare and contrast different data points, explore various scenarios, and draw their conclusions without the constant need for finance team intervention.

The adoption of the visualisation tool at Tootal Telekom is a testament to the power of technology in enhancing business efficiency and data-driven decision-making. By democratising access to data and equipping staff with the tools to explore it independently, the company fostered a more informed, agile, and proactive work environment.

Final Outcome

Project Titan's success represents a significant milestone in Tootal Telekom's journey to becoming a data-driven organisation. The transformation achieved through this project shifted the company from a backward-looking business model to one that is dynamically informed by real-time data. This change was crucial in enabling Tootal Telekom to respond swiftly and effectively to market dynamics and internal business needs.

The hallmark achievement of Project Titan was the realisation of the zero-day close. Completing the financial closing process on the last day of the month was not just about speed; it was about bringing timely, actionable insights to the forefront of business decision-making. This capability allowed the company to analyse and understand its financial position almost immediately after the month's end, significantly enhancing its strategic planning and response capabilities.

The global recognition of Project Titan across industries and geographies underscores its significance as a benchmark in financial process innovation. Achieving a zero-day close is a formidable task for any organisation, particularly for one with legacy systems and processes like Tootal Telekom.

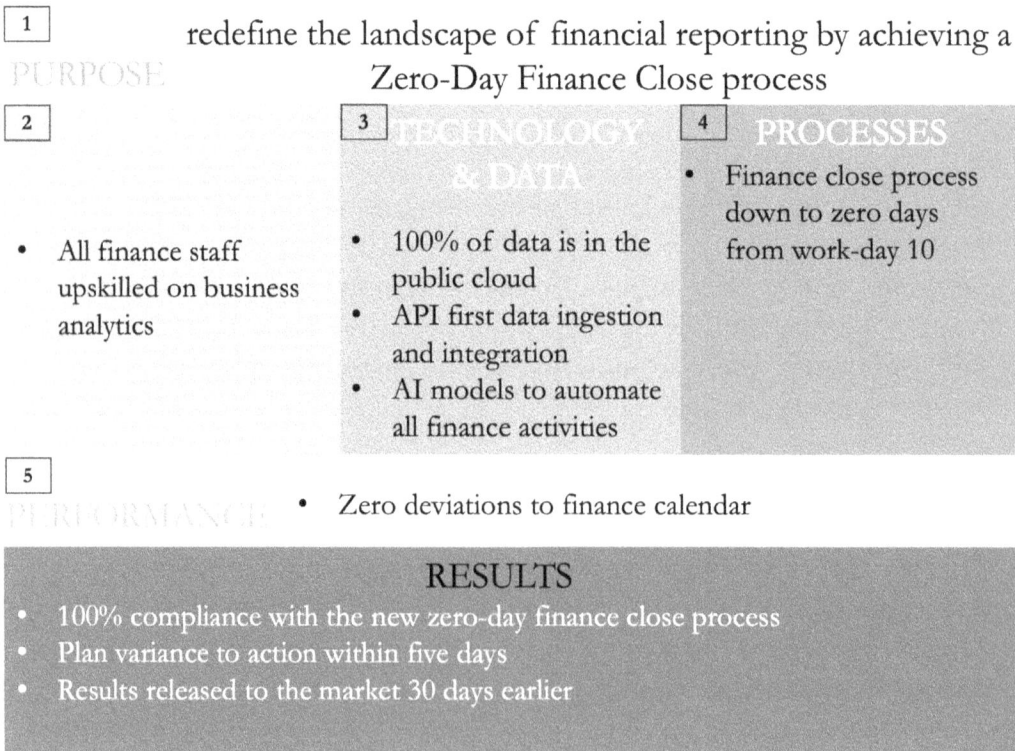

1 PURPOSE

redefine the landscape of financial reporting by achieving a Zero-Day Finance Close process

2

3 TECHNOLOGY & DATA

4 PROCESSES

- All finance staff upskilled on business analytics

- 100% of data is in the public cloud
- API first data ingestion and integration
- AI models to automate all finance activities

- Finance close process down to zero days from work-day 10

5 PERFORMANCE

- Zero deviations to finance calendar

RESULTS
- 100% compliance with the new zero-day finance close process
- Plan variance to action within five days
- Results released to the market 30 days earlier

Figure 15

Project Titan's success story serves as an inspiration and a model for other organisations striving to enhance their financial processes and overall business agility. The transformation to a data-driven organisation, as demonstrated by Tootal Telekom, highlights the importance of embracing AI, technology, and innovation in today's rapidly evolving business landscape. Figure 15 opposite outlines the strategic objectives set at the start of the transformation project.

14

Case Study 4 – Auroria Tax Authority Transforms the Country's Tax System

Names, locations, and specific financial data have been changed to protect privacy.

Background

Auroria, with its 25 million inhabitants, vast land, abundant natural resources, and highly educated workforce, has emerged as a global leader in various sectors. The country's continuous economic growth over the past three decades has led to its admission into the G20, reflecting its significant role and influence in the global economy. This growth is underpinned by its rich resources and skilled population, driving Auroria's status as a prominent and prosperous nation on the world stage.

The Auroria Tax Authority (ATA), the government's revenue collection agency, had garnered global attention for its issues with corruption, revenue leakage, and ineffective systems and processes. These shortcomings had negatively impacted the country's reputation, highlighting the need for significant reforms to address these deep-rooted problems. The situation underscored the importance of transparency, efficiency, and integrity in government operations, especially in critical areas like tax collection and administration.

Challenge

When the Commissioner of Taxation was appointed in 2019, he was tasked with modernising the organisation. The ATA had to overcome its reputation and transform into a service-oriented, data-driven organisation. This involved changing its culture, improving the staff experience, and enhancing digital and data capabilities.

The transformation was not just about technological upgrades but also about embedding better practices and simplifying taxpayer interactions. The ATA's journey involved a focus on cultural transformation, aligning with community expectations, and a commitment to ongoing improvement without an end date.

It was estimated that the ATA had a revenue leakage rate of over 60% due to significant challenges in reporting, calculations, processing, collection, and governance. This amounted to more than $25 billion a year of deficit that had to be financed through other means.

Transformation & Strategy

The ATA embarked on a transformational strategy to establish itself as a global benchmark in modern tax administration. The newly appointed commissioner worked in tandem with the Department of Treasury to secure legislative approval for this strategy. This transformative initiative represents a significant step towards rectifying the challenges faced by the ATA and enhancing its efficiency and integrity in tax administration, as illustrated in Figure 16.

To achieve its transformation goals, the ATA identified four key economy-wide changes essential for connecting everyone to its system:

1. National Identity Card Number as Tax File Number
   – This measure involved using the National Identity Card Number as the Tax File Number for individuals, streamlining identification and tax processes.

2. Bank account numbers linked to the Tax File Number – By linking bank accounts to Tax File Numbers, the ATA aimed to enhance transparency in financial transactions and improve tax compliance.

3. Business Registration Number as Tax File Number – This change meant businesses would use their registration number for tax purposes, simplifying the tax process for companies.

4. Business finance facilities linked to registration number – Linking business finance facilities to registration numbers would enable easier monitoring of business transactions and compliance.

The ATA managed to get political and economy-wide support for the above changes that were implemented across the finance sector in record time.

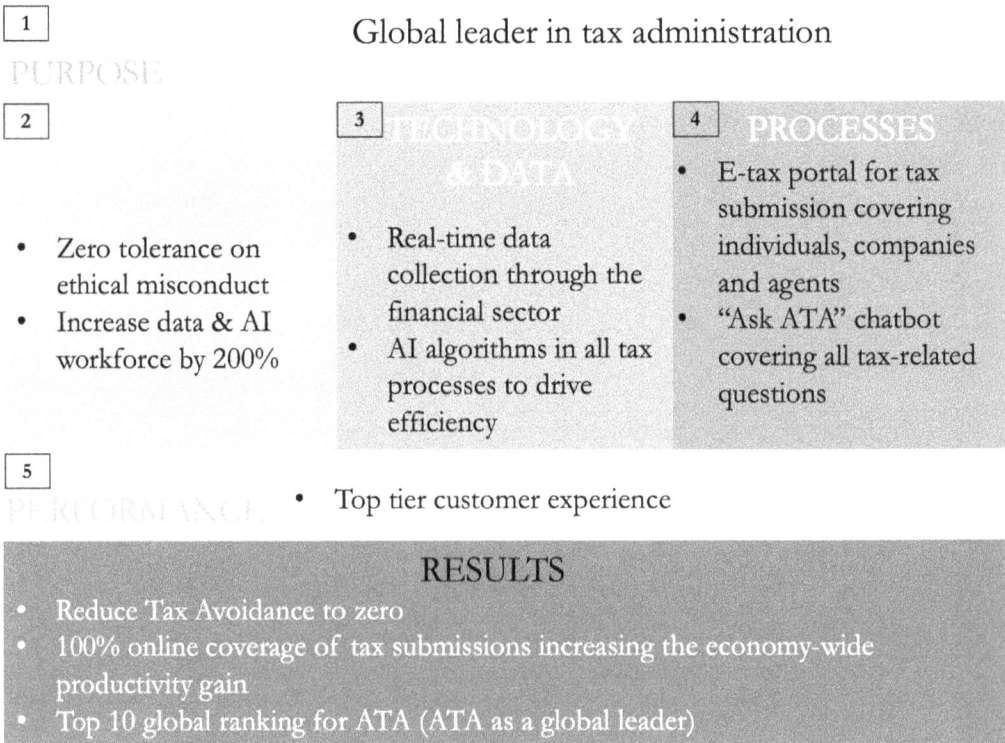

1
Global leader in tax administration

PURPOSE

2

- Zero tolerance on ethical misconduct
- Increase data & AI workforce by 200%

3 TECHNOLOGY & DATA

- Real-time data collection through the financial sector
- AI algorithms in all tax processes to drive efficiency

4 PROCESSES

- E-tax portal for tax submission covering individuals, companies and agents
- "Ask ATA" chatbot covering all tax-related questions

5
PERFORMANCE

- Top tier customer experience

RESULTS

- Reduce Tax Avoidance to zero
- 100% online coverage of tax submissions increasing the economy-wide productivity gain
- Top 10 global ranking for ATA (ATA as a global leader)

Figure 16

Connecting the Finance System

The ATA took a pivotal step by integrating the banking system with its operational framework. This integration was facilitated through the deployment of a data lake and a sophisticated cloud-based infrastructure, enabling the ATA to systematically collect and process all financial transactions occurring within the country's economy. This state-of-the-art infrastructure not only augmented the transparency of financial activities but also provided the ATA with a robust mechanism for accurately verifying tax submissions made by both individuals and businesses.

This move was more than just a technological upgrade; it represented a paradigm shift in the way financial information was managed and utilised for taxation purposes. By having real-time access to comprehensive financial data, the ATA was better positioned to detect discrepancies, prevent tax evasion, and ensure compliance.

Additionally, this approach significantly reduced the administrative burden on taxpayers, as the automated system streamlined the process of tax calculation and submission. The implementation of this cloud-based data infrastructure marked a notable stride in Auroria's journey to an efficient, transparent, and accountable financial ecosystem.

Moving to E-tax

The ATA recognised the need to modernise its approach to tax collection and management. To this end, the ATA introduced the "E-tax" system, a comprehensive digital platform designed to streamline the tax process for both individuals and businesses.

This E-tax system simplifies interactions with the tax authority, allowing taxpayers to file returns, make payments, and access their tax records online. This modern approach not only makes tax compliance more convenient but also enhances transparency and efficiency in tax collection.

For individual taxpayers, E-tax is a straightforward way to manage their tax obligations. They can easily file returns, view their tax history, and make payments online. This ease of access encourages timely and accurate tax filing, thereby improving compliance rates.

For businesses, the system's ability to handle complex tax scenarios relevant to businesses ensures that companies of all sizes can manage their tax matters effectively.

Furthermore, the ATA's adoption of a data-driven approach through the E-tax system allows for better monitoring and analysis of tax data. This capability enables the ATA to identify trends, detect non-compliance, and make informed decisions about tax policy and administration.

In summary, the E-tax system marks a transformative step in modernising the country's tax administration. It brings convenience, efficiency, and transparency to the tax process, benefiting both taxpayers and the ATA. This digital initiative positions Auroria as a forward-thinking nation in tax administration and public service delivery.

Pre-filled Returns

The ATA's introduction of pre-filled tax returns represents a groundbreaking transformation in tax administration, leveraging the seamless integration of banking data with the tax system. This innovative approach utilises advanced tax algorithms to significantly streamline the tax filing process for both individuals and companies.

The tax algorithms play a crucial role in this new system. They automatically analyse and classify financial transactions from connected banking data. Based on this analysis, the algorithms pre-populate fields in the tax returns with relevant financial information, such as income, expenses, and deductions. This process results in the preparation of a draft tax return that is almost complete, reducing the burden on taxpayers to gather and input this data manually.

For individuals and companies, this means a drastic reduction in the time and effort required for tax filing. The pre-filled returns provide a preliminary version of their tax obligations or refunds based on the financial data already available to the ATA. Taxpayers are then required to review, verify, and, if necessary, amend the details in these draft returns before submission.

This not only simplifies and speeds up the tax filing process but also enhances the accuracy of tax submissions, as the data used is directly sourced from verified financial transactions.

The ATA's initiative significantly reduces the likelihood of errors and omissions that are common in manually filled returns. This accuracy is crucial for ensuring compliance and fairness in the tax system. It also aids in speeding up the processing of returns, leading to quicker assessments and refunds.

Overall, the implementation of pre-filled tax returns by the ATA reflects a forward-thinking approach to tax administration, harnessing technology to benefit both the authority and the taxpayers. This innovation not only modernises the tax filing process but also positions Auroria as a leader in utilising technology to improve government services.

Rules Book

Part of the ATA's transformation included the development of a customised rule book for each taxpayer. This rule book, based on the taxpayer's previous three years of tax return submissions, forms the core of both the E-tax and pre-filled returns.

By analysing the tax history of each taxpayer, the system intelligently identifies specific patterns, deductions, income sources, and other relevant tax considerations unique to each taxpayer. This information is then used to create a custom rule book that directly applies to their tax situation, including their pre-filled returns.

For taxpayers, this innovation brings a new level of transparency and understanding to the tax filing process. Individuals and companies can download their personalised rule book, which allows them to better understand the tax rules and regulations that apply specifically to them. This accessibility to tailored tax information is not only educational but also empowers taxpayers to be more engaged and informed about their tax responsibilities.

Moreover, the acceptance and embracement of the rule book by taxpayers reflect a significant shift in the public's interaction with the tax system. By providing clarity and personalised guidance, the ATA has fostered a more cooperative and informed relationship between the tax authority and taxpayers.

Tax Algorithms

The ATA successfully automated various aspects of tax calculations and decision-making, leading to a more efficient and effective system.

The ATA's AI implementation included a range of sophisticated algorithms, both supervised and unsupervised, each designed to handle specific elements of the tax process. Supervised AI algorithms were employed to analyse historical tax data, learning from past submissions to improve the accuracy of tax calculations and identify common filing patterns. This approach allowed for a more refined and accurate process of pre-filling tax returns, ensuring that the data aligns closely with the taxpayer's financial history and current tax situation.

On the other hand, unsupervised AI algorithms were utilised to uncover hidden patterns and anomalies in tax data that might indicate errors, fraud, or unusual financial activities. These algorithms are capable of sifting through vast amounts of data to detect irregularities that may require further investigation, thereby enhancing the integrity and reliability of the tax system.

The implementation of these AI technologies transformed the role of the ATA staff. With routine calculations and basic decision-making processes automated, the staff was able to redirect their focus toward more strategic and complex tax administration matters. This shift not only increased operational efficiency but also allowed the ATA to allocate its human resources to areas where it could provide greater value, such as policy development, complex investigations, and improving taxpayer services.

Ask ATA

"Ask ATA" is the innovative chatbot developed by the ATA. Built on a powerful generative AI platform, the chatbot is augmented by ATA's extensive knowledge base, making it an invaluable resource for staff, taxpayers, and tax professionals seeking answers to tax-related queries.

The core strength of "Ask ATA" lies in its AI-driven learning capabilities, which are fuelled by a vast dataset encompassing a wide range of tax-related information. This dataset includes comprehensive details on tax policies, procedures, principles, rules, relevant legal cases, and various legal documents. By continuously analysing and learning from this information, the chatbot is able to provide accurate, up-to-date, and contextually relevant responses to inquiries.

One of the most innovative aspects of "Ask ATA" is its ability to evolve. As it interacts with users and processes their queries, it learns from these interactions, constantly refining its understanding and improving its responses. This dynamic learning process ensures that "Ask ATA" remains a reliable and current source of tax information, adapting to changes in tax laws and policies.

For the staff, "Ask ATA" serves as an on-demand knowledge assistant, helping them quickly access information and guidance, which enhances their efficiency and effectiveness in their roles. For taxpayers and professionals, it provides a user-friendly platform to get quick answers to their tax questions, making tax compliance more accessible and less daunting.

Overall, "Ask ATA" exemplifies how AI technology can be leveraged to enhance public service delivery, making complex and often overwhelming tax information more accessible and understandable to a wide range of users.

Rules-based Auditing

The Audit Department significantly enhanced its efficiency and effectiveness through the implementation of advanced workflow tools. These tools were designed to intelligently identify specific areas and tax files that require focused auditing and verification. This targeted approach allows the department to concentrate its resources on areas that are more prone to discrepancies or non-compliance, thereby improving the overall integrity of the tax system.

Each quarter, the Audit Department strategically zeroes in on particular aspects of tax filings, such as specific types of deductions or credits. This focused approach is crucial in minimising and eventually eliminating instances of tax reduction manipulation and avoidance. By honing in on these selected areas, the department can apply its expertise more effectively, ensuring that compliance is maintained and any potential misuse of the tax system is swiftly addressed.

One of the key strategies employed by the Audit Department is the advanced publication of their areas of focus. This transparency serves multiple purposes. Firstly, it provides the public and tax professionals with their auditing priorities, enabling them to better understand the compliance landscape and adjust their filings accordingly. Secondly, this pre-emptive disclosure acts as a deterrent against potential avoidance tactics. When taxpayers are aware that certain areas will be scrutinised closely, they are more likely to comply with the regulations from the outset.

The use of these workflow tools, coupled with the department's strategy of publicising its focus areas, underscores the ATA's commitment to fostering a transparent, fair, and compliant tax environment. It demonstrates an innovative blend of technology and strategic communication, enhancing the effectiveness of the auditing process and building a culture of compliance among taxpayers.

Final Outcome

The transformation undertaken by the ATA represents a significant milestone in public sector administration. The comprehensive changes have revolutionised its operations, establishing a robust and efficient ecosystem for revenue collection. These efforts have not only set a new benchmark for public sector experience but have also positioned the ATA as a global role model for seamless and effective tax administration.

The impact of these transformative changes extends beyond improved operational efficiency. The ATA has successfully cultivated an environment where tax compliance is more straightforward and transparent for taxpayers, which has significantly contributed to reducing tax avoidance. Achieving a near-zero level of tax avoidance is a testament to the effectiveness of its strategic initiatives, including the integration of advanced technologies, the introduction of innovative systems like pre-filled returns, "Ask ATA", and the targeted focus of its audit department.

Furthermore, the ATA's role has expanded to become a major contributor to the national budgeting process. The accuracy and efficiency of revenue collection have provided the government with a reliable and predictable fiscal foundation, essential for effective budget planning and allocation. This contribution is crucial, as it ensures that the government has the necessary resources to fund public services, infrastructure development, and other vital national projects.

In conclusion, the ATA's journey of transformation has not only achieved its primary objective of minimising tax avoidance but has also significantly enhanced the overall quality of tax administration. The authority's success story serves as an inspiring example of how strategic reforms, when coupled with technological innovation, can lead to exceptional outcomes in public sector management and service delivery.

15

Case Study 5 – Euro Assurance Transforms Its Global Auditing Process

Names, locations, and specific financial data have been changed to protect privacy.

Background

Euro Assurance, established as a tier 2 global auditing and assurance firm, has carved out a significant niche in the international auditing sector. With a presence in 50 locations worldwide, the company has built a reputation for delivering robust and reliable auditing and assurance services across a diverse range of industries. This expansive network not only signifies its global reach but also demonstrates its capability to cater to a wide spectrum of clients, from burgeoning startups to established multinational corporations. Over the years, Euro Assurance has consistently emphasised quality and precision in its services, ensuring compliance with international standards and adapting to the varying regulatory environments of its operational territories.

Recently, Euro Assurance embarked on a transformative journey under the leadership of its newly appointed global head of assurance. Recognising the dynamic nature of the global financial landscape and the growing competition from the Big 4 audit firms, this strategic move was aimed at overhauling the company's assurance processes.

The goal was to modernise its systems comprehensively, integrating cutting-edge technology and innovative methodologies. This initiative was not just about keeping pace with industry standards; it was about setting new ones. By investing in advanced tools like AI and data analytics, Euro Assurance aimed to enhance the efficiency, accuracy, and value of its services. This ambitious endeavour positioned Euro Assurance not just as a participant in the global auditing and assurance industry but as a potential challenger to the dominance of the Big 4, promising to bring fresh perspectives and competitive services to the market.

Challenge

The global review conducted on Euro Assurance practices brought to light that the planning and execution of these audits were heavily reliant on the individual experience and prior knowledge of the partners and their staff. This personalised approach, while drawing on the expertise of its professionals, led to considerable variability in the auditing process. Such differences were especially pronounced across the various geographical locations where EA operated, indicating a lack of a standardised methodology.

This finding underscored the necessity for EA to develop and implement a more robust, standardised process for audits and assurance services. The aim would be to ensure consistency and uniformity across all regions, which would also improve the overall effectiveness and reliability of EA's services.

By systematising the auditing process, EA could assure clients and stakeholders of the same high level of service and scrutiny, regardless of location. Such a move would also facilitate better compliance with international standards and regulations.

Vision

Euro Assurance was keenly focused on developing solutions that address the specific gaps identified in their recent review. The company's vision was to overcome challenges such as the inconsistent approach in auditing processes, reliance on manual procedures, and inadequate record-keeping. To tackle these issues, Euro Assurance considered a multifaceted strategy that included the integration of technology, process reengineering, and enhanced training programs.

1. Technology integration – This included the use of AI and data analytics to ensure consistency in audit methodologies across different locations. Automation not only reduced the reliance on manual processes but also enhanced accuracy and efficiency.

2. Process standardisation – This involved creating a comprehensive framework that guides all aspects of the auditing process, ensuring that audits are conducted consistently, irrespective of the region or personnel involved.

3. Improving record-keeping and data management – These systems would ensure meticulous record-keeping, allowing for better tracking of audit trails and easier access to historical data. Enhanced record-keeping would also aid regulatory compliance and offer greater transparency to clients.

4. Staff training and development – Recognising that technology alone was not a panacea, Euro Assurance committed to investing in continuous training and development programs for its staff. This included education on new technologies, standard processes, and best practices in auditing and assurance. By equipping its workforce with the necessary skills and knowledge, the firm ensured that its human capital is as advanced as its technological resources.

Through these initiatives, Euro Assurance was positioned to not only rectify the shortcomings but also to elevate its service offerings, aligning with the best practices in the industry. This comprehensive approach was expected to enhance the overall quality and reliability of the company's auditing and assurance services, solidifying its position as a formidable entity in the global market. Figure 17 below depicts the firm's vision.

VISION — Global audit process that connects all clients and staff via a platform that standardises practices globally

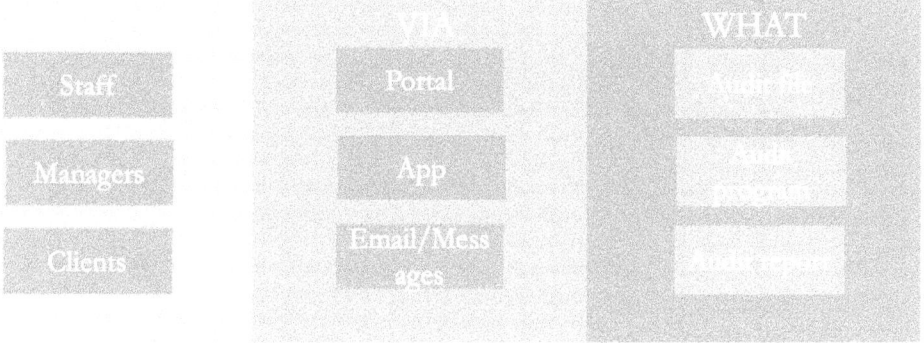

VIA — WHAT

Staff — Portal — Audit file

Managers — App — Audit program

Clients — Email/Messages — Audit report

UNDERPINNED BS

- Fully automated workflow that generates a risk-based audit program
- AI-driven algorithms that identify areas for audit
- Access is provided to everyone who sees information in real-time
- Consistent audit approach globally
- Digital audit trail that captures notes, comments, feedback, reports and emails

Figure 17

Large Investment

Euro Assurance made a significant commitment to modernising its assurance technology infrastructure with a substantial investment of $200 million over a three-year period. This investment was strategically allocated to develop and enhance three broader areas: a data lake, an AI ecosystem, and a digitised workflow with web and app interfaces.

1. Data lake development – The creation of a data lake was a pivotal part of the company's technological overhaul. This decentralised repository allowed the firm to store vast amounts of structured and unstructured data from various sources in local, accessible locations in the public cloud. The data lake enabled more efficient data management and analysis, providing a robust foundation for data-driven decision-making and auditing processes. This investment

significantly improved data accessibility and integrity, which is crucial for comprehensive audits and assurance services.

2. AI ecosystem implementation – This ecosystem facilitated advanced data analytics, predictive modelling, and automated processes, significantly enhancing the efficiency and accuracy of auditing and assurance services. The AI ecosystem revolutionised how the company handles complex datasets, identifies patterns, and derives insights, thereby elevating the quality of its services to a new standard.

3. Digitised workflow with web and app interface – The third focus area was the digitisation of workflow processes complemented by user-friendly web and app interfaces. This initiative aimed to streamline auditing and assurance processes, making them more efficient and accessible to clients and staff. The introduction of digital tools and platforms facilitated real-time collaboration, reporting, and communication. This not only improved operational efficiency but also enhanced the client experience, providing them with easy access to information and services through intuitive digital channels.

The company's investment in these three key areas demonstrated a commitment to embracing digital transformation and innovation. By doing so, the firm was not only addressing technological gaps but was also positioning itself as a leader in the digital future of auditing and assurance services. This strategy was expected to significantly boost the company's competitive edge in the global market, aligning its operations with the emerging trends and demands of the digital era.

Global Data Lake with Local Hosting

Euro Assurance's global presence brought with it the challenge of complying with various local regulatory and data sovereignty requirements. To effectively navigate these complexities, the company adopted a country-specific data hosting approach. This strategy ensured that the firm adhered to the unique data protection laws and regulations of each country in which it operated.

1. Country-specific data hosting – These data-specific repositories were tailored to meet the individual legal and regulatory requirements regarding data storage and management in each jurisdiction. This approach also addressed concerns related to data sovereignty, which is crucial for maintaining trust with local clients and regulatory bodies.

2. Uploading of historical data – Each country within the company's network was responsible for uploading its historical data to its respective data lake. This process involved transferring years of auditing and assurance records into the newly established local data repositories. This meticulous migration was crucial for maintaining a comprehensive and accessible historical record, enhancing the quality and depth of audit services.

3. Secure private network connection – To ensure both the integrity and security of data across its global operations, the company connected these country-specific data lakes through a secure private network. This network facilitated safe and efficient data transfer and access between different countries' data repositories while strictly adhering to cybersecurity protocols.

4. Cybersecurity measures – Recognising the critical importance of data security, especially in the highly sensitive auditing and assurance industry, the company implemented robust cybersecurity measures. These included advanced encryption, regular security audits, intrusion detection systems, and continuous monitoring to safeguard against cyber threats.

By prioritising local compliance and data security, Euro Assurance was able to offer its services across different regions while maintaining the highest standards of data integrity and protection. This careful balance of localisation and global connectivity positioned the firm as a trusted and reliable partner in the international auditing and assurance sector.

AI is the Brain Behind the Stack

Euro Assurance's adoption of AI in its global auditing process signified a groundbreaking shift in how auditing and assurance services were delivered. At the core of this technological advancement was a centralised decision engine powered by AI, which fundamentally transformed the auditing process.

1. Centralised AI decision engine – This served as the hub for its global audit operations. The AI algorithms analysed historical data and reports, identifying patterns and anomalies that might indicate audit risks.

2. Automated risk-based planning – One of the key capabilities of the AI engine was its ability to automatically generate risk-based audit plans. By evaluating vast quantities of data, the AI identified potential risk areas and allocated resources accordingly. This automated planning allowed the staff to focus on critical aspects of the auditing process, enhancing efficiency and effectiveness.

3. Combination of supervised and unsupervised learning – The AI engine employed both supervised and unsupervised learning algorithms. Supervised learning algorithms were trained on historical data with known outcomes to predict future events, while unsupervised learning algorithms detected patterns and relationships in data without prior training. This combination enables the AI to continuously learn and adapt, refining its audit strategies over time.

4. Learning from global data for local application – A significant advantage of the system was its ability to learn from data on a global scale and then apply these insights locally. When the AI identified specific trends or risks in one location, it leveraged these findings to enhance audits in new locations. This capability facilitated the transfer of knowledge and best practices across borders, ensuring that each local office benefited from global insights.

Euro Assurance's integration of AI into its auditing process represented a significant leap forward in the field of auditing and assurance. By leveraging advanced AI technologies, the company was able to offer more accurate, efficient, and globally informed audit services.

Interconnectedness

Euro Assurance's innovative platform is a comprehensive ecosystem intricately designed to connect staff, clients, key personnel, and specific members of regulatory bodies, all tailored to meet their unique access needs. This system revolutionised the traditional audit engagement process, transforming it into a dynamic, digital marketplace for all audit-related activities.

1. Interconnected digital platform – The platform acts as a central hub where all parties involved in the auditing process can interact seamlessly. This interconnectedness ensures that communication, data sharing, and collaboration are streamlined, enhancing the efficiency and effectiveness of the auditing process.

2. Customised access control – This intelligently restricts access based on roles and responsibilities. Clients can view and interact with information pertaining exclusively to their organisation, ensuring confidentiality and relevance. Meanwhile, staff members have access to a broader range of data, including audit programs, past audit files, queries, questions, and communication channels with clients.

3. Client-specific access – The platform was designed to provide clients access only to information that is relevant to them. This not only simplifies the user experience for clients but also reinforces the security and privacy of sensitive audit information. Clients are granted visibility to their own audits without the risk of exposure to other clients' data.

4. Comprehensive staff access – The staff, on the other hand, are granted a comprehensive access level. This includes detailed audit plans, historical audit data, ongoing queries, and communications, allowing them to manage and conduct audits effectively and efficiently. Such access empowers the staff to perform their duties with greater insight and context.

5. Robust identity and authentication controls – These controls were designed to verify the identities of all users and ensure that each individual can only access the information and tools appropriate for their role. These controls include multifactor authentication and time-sensitive passcodes. This is crucial for maintaining the integrity of the auditing process and protecting sensitive information.

6. Secure and compliant infrastructure: The platform's infrastructure was built with a strong emphasis on security and compliance with industry standards and regulations such as FIDO and Essential Eight.

By offering a secure, role-specific, and interconnected environment, Euro Assurance enhanced the auditing experience for all parties involved. This platform not only increased operational efficiency and data security but also set a new standard in client engagement and regulatory compliance within the audit industry.

App and Web Interface

The app and web interface of Euro Assurance's platform significantly enhanced the audit engagement process by introducing a level of convenience and efficiency previously unattainable. This digital interface, accessible both through web and mobile applications, was meticulously designed to cater to the diverse needs of auditing processes.

1. Mobile app for on-the-go data entry – The app allows auditors to upload data, scanned documents, and pictures directly into the system, providing real-time support for audit evidence even when they are in the field. This feature is particularly useful for documenting physical evidence and ensures that the audit records are comprehensive and up-to-date.

2. Facilitating physical counts of inventories and assets – The app is instrumental in conducting physical counts of inventories and assets, which are integral aspects of the auditing process. Auditors can use the app to record counts, take pictures, and even tag locations, thereby streamlining the inventory and asset verification process.

3. Mobile camera integration – Another innovative feature of the app is its ability to capture photos and videos, which can be used for recording audit meetings or documenting physical assets and inventory.

4. Seamless integration with meeting apps – Recognising the importance of remote communication, especially in a post-pandemic world, the platform seamlessly integrated with popular meeting applications like Zoom and Microsoft Teams. This integration facilitates virtual meetings, discussions, and collaborations among audit teams, clients, and other stakeholders. It ensures that even when in-person meetings are not possible, the auditing process can continue without interruption.

5. User-friendly web interface – In addition to the mobile app, the platform's web interface provides a more expansive view suitable for in-depth analysis, planning,

and report generation. The web interface was designed to be intuitive and user-friendly, allowing easy navigation and access to a wide range of functionalities necessary for comprehensive audit management.

The combination of the app and web interface on Euro Assurance's platform brought a new level of agility and flexibility to the auditing process. By leveraging these digital tools, the company is able to enhance the efficiency, accuracy, and convenience of audit engagements, offering a superior experience to both auditors and clients.

Final Outcome

The introduction of Euro Assurance's new audit platform signified a pivotal moment both for the company and for the global auditing sector. This transformation underscored a profound shift towards embracing cutting-edge technology and AI in the field of auditing, marking a significant milestone in EA's corporate journey.

1. Revolutionising auditing practices – By integrating AI, advanced data analytics, and digital tools, the platform not only automates and streamlines traditional auditing processes but also introduces new levels of accuracy, efficiency, and insight. This shift from conventional, often manual auditing methods to a more sophisticated, technology-driven approach is a clear indication of the firm's commitment to innovation and excellence.

2. Setting new industry standards – The platform's capabilities, such as real-time data processing, AI-driven risk assessment, and seamless integration with digital communication tools, demonstrate what the future of auditing can look like. It challenges other firms in the sector to rethink and upgrade their own processes, thereby elevating industry standards as a whole.

3. Global impact and influence – The global reach of Euro Assurance means that this change extends well beyond its own operations. As a firm with a presence in multiple countries, the adoption of such an advanced platform has the potential to influence auditing practices worldwide. It showcases how technology can be leveraged to meet diverse regulatory requirements and client needs across different markets, setting a precedent for others to follow.

4. Enhancing transparency and trust – The advanced capabilities of the platform, particularly in terms of data management and security, play a crucial role in enhancing transparency and trust in the auditing process. By providing more reliable and comprehensive auditing services, the firm contributes to higher standards of corporate governance and financial reporting globally.

5. Responding to evolving business environments – This move by Euro Assurance was also a response to the rapidly evolving business environments and regulatory landscapes. In an era where data is gold, and compliance requirements are increasingly stringent, the company's changes enabled it to better meet these challenges and cater to the evolving needs of clients.

Euro Assurance's new audit platform was more than just an upgrade; it was a bold and strategic move into a new era of auditing. It reinforced the company's position as a forward-thinking leader in the industry as well as paving the way for broader transformations in how auditing and assurance services are delivered globally.

16

Case Study 6 – KNN Bank Overhauls Its Risk Platform

Names, locations, and specific financial data have been changed to protect privacy.

Background

KNN Bank is one of the largest banks in Australia and a significant player in the banking sector across the Asia-Pacific region. Founded in the mid-20th century, KNN has a rich history that spans over 70 years. Over the decades, KNN has grown substantially, both organically and through a series of strategic acquisitions, solidifying its position as a leading banking and financial services group. The bank has played a pivotal role in the economic development of Australia and New Zealand, financing trade and supporting businesses and individuals across the region.

KNN's expansion into the Asia-Pacific region has been a key part of its growth strategy, making it one of the few Australian banks with an extensive network in Asia, the Pacific Islands, Europe, the US, and the Middle East. The bank offers a wide range of financial services, including retail and commercial banking, wealth management, and investment banking.

KNN is known for its innovation in the banking industry, having been an early adopter of computer technology in banking operations and continuously evolving its digital services to improve customer experience. The bank's commitment to sustainability and responsible banking practices has also been a notable aspect of its operations. KNN's focus on diversity, community engagement, and environmental sustainability reflects its aim to not only be a leader in financial services but also to impact the communities it serves positively.

Challenge

KNN, like many large financial institutions, faced a range of challenges in risk management stemming from both internal operational complexities and external market dynamics. One of the primary challenges was navigating the ever-changing regulatory landscape. As a bank operating in multiple countries across the Asia-Pacific region and beyond, KNN must comply with a diverse array of regulatory standards and practices. These regulations continually evolve, especially in response to the global financial landscape and emerging risks in cybersecurity and data protection.

Ensuring compliance while managing operational risks associated with cross-border transactions and diverse market practices added layers of complexity to KNN's risk management strategy. Additionally, the increasing focus on anti-money laundering (AML) and counter-terrorism financing (CTF) regulations required KNN to constantly update its systems and processes to detect and prevent illicit activities, adding further operational demands.

Another significant challenge for KNN in risk management was the rapid advancement of technology and the accompanying cyber threats. As digital banking became more prevalent, the risk of cyberattacks and data breaches escalated, posing a threat to the security of customer data and the integrity of banking systems. Managing this cyber risk was a constant battle, requiring ongoing investment in advanced cybersecurity measures and regular updates to IT infrastructure.

Furthermore, the bank needed to manage the risks associated with adopting new technologies and digital platforms, balancing the need for innovation with the potential operational and reputational risks. The rise of fintech and digital-only banking competitors also presented strategic risks. KNN must continuously evolve to meet changing customer expectations and maintain its competitive edge in a rapidly transforming digital landscape. These challenges underscored the need for a robust, proactive risk management framework that could adapt to both the external market environment and internal operational shifts.

An in-depth review of KNN's framework, although comprehensive, disclosed that thousands of unmitigated risks had accumulated, primarily due to the absence of a centralised system to manage them effectively. This oversight in the risk management architecture signified a major vulnerability, as it hindered the bank's ability to track, prioritise, and address these risks systematically.

The decentralised nature of the system led to inconsistencies in risk evaluation and mitigation strategies across different departments and geographical locations. This fragmentation not only posed a threat to the operational stability of the bank but also raised concerns among stakeholders about the bank's ability to manage complex risk scenarios efficiently.

As a result, KNN had to overhaul its risk management, with a focus on developing a more integrated, centralised system that can effectively identify, assess, and mitigate risks on a comprehensive scale.

Risk Management Framework

KNN's risk management framework (RMF) was a comprehensive and sophisticated system designed to identify, assess, and manage the various risks that the bank faced in its operations. This framework was central to KNN's strategy, reflecting its commitment to maintaining financial stability and safeguarding the interests of its stakeholders, including customers, employees, and shareholders. Figure 18 below outlines KNN's risk management framework.

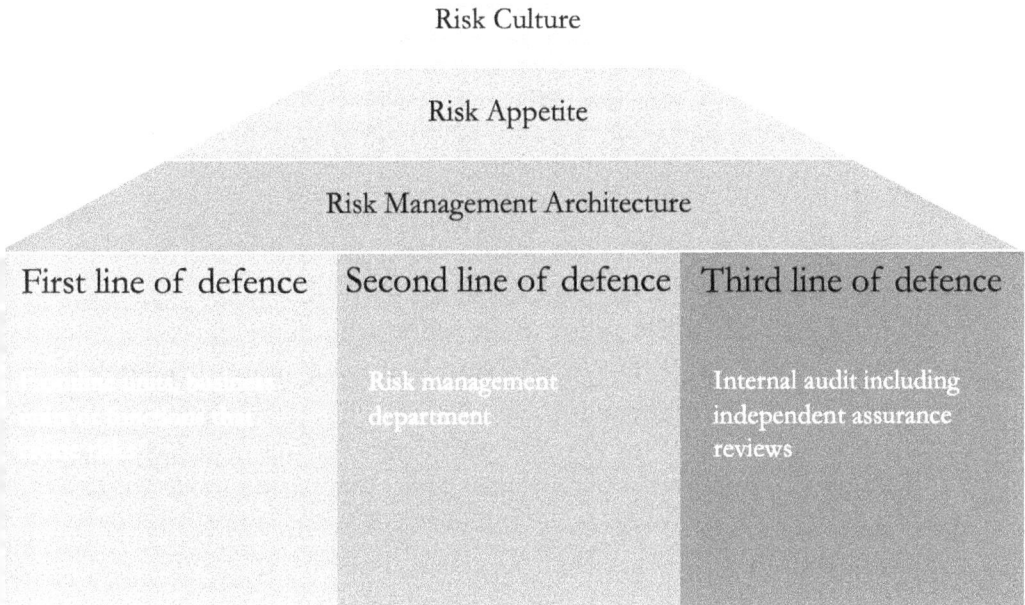

Risk Culture

Risk Appetite

Risk Management Architecture

First line of defence	Second line of defence	Third line of defence
	Risk management department	Internal audit including independent assurance reviews

Figure 18

Risk culture refers to the collective attitude, values, and behaviours related to risk awareness, risk-taking, and risk management within an organisation.

Risk appetite represents the level of risk that an organisation is willing to accept in pursuit of its strategic objectives.

Overhauling the Risk Management Platform

The overhaul of KNN's risk management platform was driven by the recognition of the need for a more cohesive and streamlined approach to managing the diverse risks the bank faces. The new platform was envisioned to serve as a unified hub for risk identification, assessment, monitoring, and mitigation across the entire organisation. This consolidation was crucial for eliminating the silos that previously characterised KNN's risk management practices, allowing for a more coherent and comprehensive view of risk at both the micro and macro levels.

By bringing all risk-related data and processes onto one platform, KNN aimed to enhance the accuracy and efficiency of its risk management, ensuring that risks are identified quickly and addressed proactively. The platform leveraged advanced analytics, AI, and ML to provide real-time insights and predictive risk assessments, significantly improving the bank's ability to respond to dynamic risk environments.

The revamped platform was also a response to the evolving regulatory landscape and the increasing complexity of financial markets. It enabled KNN to better comply with diverse regulatory requirements across different regions, ensuring more consistent adherence to global standards.

Furthermore, the centralisation of risk management activities better facilitated communication and collaboration among different departments and regions within the bank. This integration was not just a technical upgrade; it was also a cultural shift towards a more unified and collaborative approach to risk management. Training and development would be integral to the implementation of this new system, ensuring that all employees are equipped with the necessary skills and understanding to utilise the platform effectively. Through this overhaul, KNN positioned itself not just to manage risks more effectively but also to harness them as opportunities for growth and innovation.

Key Features of the Real-time Risk Management Platform

KNN's new centralised, automated risk management system was a state-of-the-art solution designed to enhance the bank's ability to manage risks effectively and efficiently. This system incorporated several key features that mark a significant advancement in KNN's approach to risk management:

1. All risks quantified – One of the primary requirements of the new RMS was that all risks have to be quantified and validated by the respective risk owner and divisional risk team. This enables the rolling up of

risk exposure at the group level for decision-making, including managing the overall risk appetite.

2. Centralised risk database – This database consolidates risk-related data across the entire organisation. This unified repository facilitates comprehensive risk exposures and better data analysis, enabling more informed decision-making.

3. Automation and AI integration – The system leveraged automation and AI to streamline risk assessment and monitoring processes. AI algorithms analyse vast amounts of data to identify potential risks, predict future trends, and suggest mitigation strategies. This reduces the reliance on manual processes and enhances the speed and accuracy of risk management activities.

4. Real-time monitoring and reporting – The platform was equipped with real-time monitoring capabilities, enabling the bank to track risk exposures as they evolve. This feature is crucial for timely responses to emerging risks. The system also provides automated reporting tools, making it easier to communicate risk statuses to stakeholders and regulatory bodies.

5. Compliance management tools – The system includes tools designed to ensure adherence to relevant laws and regulations. These tools are regularly updated to help KNN stay compliant in different jurisdictions.

6. Customisable risk assessment frameworks – The system allows for the customisation of risk assessment frameworks to suit the specific needs of different business units within the bank. This flexibility ensures that risk assessments are relevant and effective for each area of operation.

7. Integrated risk mitigation strategies – In addition to risk identification and assessment, the system provides integrated tools for risk mitigation, such as the ability

to plan, implement, and track the effectiveness of risk mitigation activities.

8. User-friendly interface and accessibility – Despite its complexity, the system was designed with a user-friendly interface, making it accessible to all relevant staff members. This ease of use encouraged widespread adoption and effective utilisation of the system across the bank.

9. Scalability and adaptability – The system was built to be scalable, allowing it to adapt to the growing and changing needs of the bank, ensuring that it remains a valuable tool for risk management in the long term.

10. Enhanced cybersecurity measures – Given the sensitivity of risk-related data, the system includes robust cybersecurity measures such as multifactor authentication to protect against unauthorised access and data breaches.

In short, KNN's new risk management system is a comprehensive and technologically advanced platform that addresses the multifaceted nature of risk in the banking sector.

How AI is at the Centre of the new RMS

The integration of AI technologies into the system was a strategic move aimed at leveraging the latest advancements in ML and data analytics to enhance the bank's risk management capabilities.

1. Advanced risk detection and analysis – AI algorithms were employed to sift through vast amounts of financial data, identifying patterns and anomalies that might indicate potential risks. This capability allows for early detection of issues that human analysts might overlook, enabling proactive risk management. The AI system can analyse historical data to predict future risk scenarios, providing valuable foresight in a rapidly changing financial landscape.

2. Predictive analytics and risk forecasting – By utilising historical data, current market trends, and various risk indicators, the AI system can forecast potential risks before they materialise. This predictive capability is crucial for KNN in strategising and preparing for various market conditions and operational challenges.

3. Automated compliance monitoring – The system was programmed to stay updated with the latest regulatory changes across different jurisdictions. It automatically reviews transactions and operational processes for compliance, flagging any potential issues for further review. This automation greatly reduces the manual workload and minimises the chances of compliance breaches.

4. Enhanced decision-making – AI-driven insights and analytics provide risk management teams with sophisticated tools to make informed decisions. The system offers recommendations based on data-driven analysis, enabling the teams to weigh the risks and benefits of various decisions with greater precision.

5. Customisation and learning – The AI system was designed to learn and adapt over time. It can be customised to the specific needs and risk profiles of different segments of the bank's operations, ensuring that the risk management strategies are always aligned with the unique aspects of each business unit.

6. Efficiency in risk management processes – By automating routine and repetitive tasks, AI allows the bank's staff to focus on more strategic aspects of risk management. This shift not only improves operational efficiency but also enhances the overall quality and effectiveness of the risk management function.

7. Cybersecurity and data protection – In addition to risk management, AI is also integral to cybersecurity efforts within the system. It helps monitor network

activities, detect potential cyber threats, and respond to security incidents in real-time.

AI was not just an addition to KNN's risk management system; it was a transformational force that redefined the entire approach to managing risk. With AI at its centre, KNN's risk management system is more agile, insightful, and effective, positioning the bank better to navigate the complexities of the modern financial world.

AI Algorithms and Risk Categories

KNN uses AI algorithms extensively in detecting, preventing, monitoring, and managing risks across the banks. A few areas of KNN's approach are outlined below.

1. Capital adequacy risk – The AI system facilitates sophisticated risk-weighted asset calculations by processing and analysing large volumes of data related to the bank's asset portfolio, classifying assets based on risk exposure. This level of analysis is pivotal in determining the bank's required capital reserves, ensuring compliance with Basel III and IV regulatory standards. Additionally, AI predictive modelling capabilities are utilised in credit risk assessment, enabling the bank to anticipate potential defaults and adjust capital buffers accordingly. AI also plays a crucial role in stress testing and scenario analysis, allowing KNN to simulate various adverse economic conditions and gauge their impact on capital adequacy.

2. Credit risk – These AI models analyse vast quantities of data, including transaction histories, market trends, and economic indicators, to identify patterns and potential risks in borrower profiles. This approach allows for more nuanced and personalised risk assessments, leading to better-informed lending decisions. AI predictive capabilities are particularly valuable in foreseeing potential defaults and assessing the likelihood of repayment, thus enabling KNN

to tailor its credit portfolios with a more precise understanding of risk exposures. Additionally, AI tools assist in real-time monitoring of loan performances and alerting to any signs of deteriorating credit quality. This proactive monitoring helps in the early identification and mitigation of credit risks, thereby safeguarding the bank's financial health.

3. Liquidity & funding risk – AI also assists in monitoring market liquidity indicators and analysing customer deposit behaviours, enabling the bank to anticipate and prepare for potential funding challenges. Furthermore, AI-driven tools can simulate stress scenarios, such as market downturns or sudden shifts in customer behaviour, allowing KNN to assess the resilience of its liquidity position.

4. Market risk – AI technologies enable KNN to conduct sophisticated market analyses, tracking and predicting market trends, interest rate fluctuations, and foreign exchange movements. By harnessing AI's powerful predictive analytics, KNN can anticipate market changes and make informed strategic decisions to hedge against potential risks. The AI models are trained on vast datasets, encompassing historical market data and current global economic indicators, allowing the bank to identify correlations and causations that might impact its investment portfolio. This deep insight assists KNN in optimising asset allocation, diversifying investments, and setting appropriate risk limits. Additionally, AI-driven scenario analysis and stress testing are integral to KNN's approach, enabling the bank to evaluate its market risk exposure under various hypothetical conditions, such as economic downturns or geopolitical events.

5. Conduct risk – KNN integrates AI into its strategy for managing conduct risk, an area increasingly vital in the banking sector, focused on ensuring ethical,

compliant, and customer-centric business practices. AI aids KNN in monitoring and analysing a vast array of transactions and interactions, flagging potential instances of misconduct, such as unethical sales practices, breaches of compliance, or deviations from standard operational procedures. By leveraging natural language processing and ML algorithms, the AI system can scrutinise communication channels for any signs of non-compliant behaviour or customer mistreatment. This level of oversight is crucial in proactively identifying and addressing issues that could lead to reputational damage or regulatory penalties. Additionally, AI predictive capabilities assist in identifying patterns or trends that may indicate systemic risks in conduct, enabling the bank to implement corrective measures before issues escalate. Beyond detection, AI also plays a role in training and awareness, helping to disseminate conduct risk management policies throughout the organisation and ensuring that employees are cognizant of the expected standards.

Final Outcome

The implementation of a real-time centralised risk management system marked a significant stride for KNN in its ambition to become a global leader in risk management within the banking sector. The early indications of the system's effectiveness were highly encouraging, showcasing the transformative impact of integrating AI and digital technologies into the bank's risk management framework.

One of the most notable achievements of this new system was the tools given to the Chief Risk Officer (CRO). With just a click of a button, the CRO could now access a comprehensive view of the group's risk profile.

This instant accessibility to in-depth risk insights is a testament to the power of AI in synthesising and analysing vast amounts of data from various sources within the bank. The system's ability to provide real-time data plays a crucial role in enabling proactive risk management decisions. It allows the bank to quickly adapt to changing market conditions, regulatory requirements, and internal risk landscapes. Moreover, this centralisation of risk data fosters greater transparency and consistency in risk assessment across the bank's global operations.

The implementation of this system was not just a technological upgrade; it was a paradigm shift in how risk is managed in the banking industry. By leveraging AI and digital technologies, KNN is set a new standard for risk management, characterised by speed, accuracy, and strategic foresight. This positions KNN at the forefront of the industry, ready to tackle the challenges of an increasingly complex and dynamic financial world and solidify its reputation as an innovator and leader in risk management.

17

Ethical Considerations and Legislative Landscape

Ethical Considerations

The integration of AI into various sectors has raised numerous ethical considerations that are essential to address for the responsible development and deployment of these technologies. These ethical concerns span a wide range of issues, from privacy and data security to fairness and accountability.

1. Privacy and data security

One of the foremost ethical concerns in AI is the handling of personal and sensitive data. AI systems often require vast amounts of data to learn and make decisions. This raises questions about what data and how that data is collected, stored, and used. Ensuring privacy means safeguarding personal information from unauthorised access or breaches.

Moreover, there is a growing concern about AI's potential to carry out surveillance and infringe upon individual privacy rights. Ethical AI deployment necessitates stringent data protection measures and respect for user privacy, aligning with legal standards like the EU's General Data Protection Regulation (GDPR).

2. Bias and fairness

AI systems are only as unbiased as the data they are trained on and the algorithms that drive them. There is a risk that AI can perpetuate and amplify existing biases, leading to unfair outcomes, especially in areas like hiring, law enforcement, and lending.

Ethical AI requires a conscious effort to identify, mitigate, and monitor biases. This involves diverse and representative datasets for training, transparent algorithmic processes, and continuous oversight to ensure decisions are fair and non-discriminatory.

3. Transparency

Another ethical challenge is the "black box" nature of some AI systems, where decision-making processes are unclear and not easily understood by humans. This lack of transparency can lead to trust issues, especially in critical applications like healthcare or criminal justice.

Ethical AI should be transparent and explainable, meaning that its decisions can be understood and interpreted by human users. This is crucial for building trust, accountability, and understanding the rationale behind AI decisions.

4. Accountability and responsibility

Determining accountability in decisions made by AI systems is complex. In cases where AI leads to unintended or harmful outcomes, it is challenging to assign responsibility, especially when multiple entities such as developers, users, and operators are involved.

Ethical AI requires the establishment of guidelines and standards for AI development and usage, ensuring that there are mechanisms in place to hold entities such as developers and operators accountable for the AI's actions or decisions.

5. Impact on employment and society

AI's impact on the workforce and societal structures is a significant ethical consideration. The automation of tasks previously done by humans raises concerns about job displacement and the widening of economic inequality.

Ethical AI integration into the workforce involves considering the societal and economic impacts, providing retraining and education programs for those affected, and ensuring that AI augments rather than replaces human capabilities. Additionally, there should be a focus on developing AI in ways that benefit society as a whole, addressing societal challenges and enhancing the quality of life.

In conclusion, as AI continues to advance and integrate into various facets of life, addressing these ethical considerations becomes paramount. This requires a collaborative effort from technologists, ethicists, policymakers, and stakeholders to ensure that AI is developed and used in ways that are responsible, fair, and beneficial for society.

AI Legislation in the EU

The European Union (EU) has been at the forefront of developing comprehensive legislation to regulate the use and deployment of AI. Recognising the potential impacts and challenges posed by AI, the EU has sought to create a legal framework that balances innovation with fundamental rights and safety.

In April 2021, the European Commission proposed the first-ever legal framework on AI, known as the *AI Act*. This proposed legislation is part of a broader European approach to digital transformation and aims to ensure that AI across the EU is safe and respects EU laws and values.

The act categorises AI systems based on the risk they pose, from "unacceptable risk" to "low risk", with corresponding regulatory requirements. On 8 December 2023, the European Parliament and Council reached an agreement on the *AI Act*, providing legal certainty for innovation and growth.

1. High-risk AI systems

A significant focus of the *AI Act* is on "high-risk" AI systems, which are subject to strict obligations before they can be put on the market. These obligations include adequate risk assessment and mitigation systems, high-quality datasets to train the AI (to minimise discriminatory outcomes), logging of activity to ensure traceability of results, detailed documentation providing all information necessary on the system and its purpose, clear and adequate information to the user, and appropriate human oversight.

2. Prohibited AI practices

The proposed legislation also identifies certain AI practices as unacceptable and prohibits them in the EU. This includes AI systems that deploy subliminal techniques or exploit vulnerabilities of specific groups of persons due to their age or physical or mental disability, which can cause physical or psychological harm. Also, AI systems used for social scoring by governments are banned.

3. Transparency rules for certain AI systems

For AI systems that interact with humans (e.g., chatbots), the legislation mandates transparency so that users know they are interacting with a machine. Similarly, AI-enabled products used for purposes like emotion recognition or biometric categorisation are subject to strict requirements.

4. Enforcement and penalties

Penalties can be steep, with fines of up to 7% of total worldwide annual turnover for companies that violate prohibitions on certain AI practices.

The EU's approach to AI legislation is one of the most comprehensive globally, reflecting its commitment to ensuring that AI development and deployment are aligned with its values of human rights, transparency, and democracy. This legislative framework is expected to set a precedent for other regions and countries, balancing the need for innovation with ethical and societal considerations.

Executive Order of President Biden on AI

US President Biden issued an executive order on 30 October 2023 that aimed at ensuring America's leadership in AI, focusing on managing AI's potential risks while harnessing its promise.

Executive Order 14110 establishes rigorous new standards for AI safety and security, prioritises Americans' privacy, advances equity and civil rights, supports consumers and workers, fosters innovation and competition, and reinforces American leadership globally. Building on previous initiatives, it calls for the following actions:

1. Require developers of the most powerful AI systems to share their safety test results and other critical information with the US government.

2. Develop standards, tools, and tests to help ensure AI systems are safe, secure, and trustworthy.

3. Protect against the risks of using AI to engineer dangerous biological materials.

4. Protect Americans from AI-enabled fraud and deception by establishing standards and best practices for detecting AI-generated content and authenticating official content.

5. Establish an advanced cybersecurity program to develop AI tools to find and fix vulnerabilities in critical software.

6. Order the development of a National Security Memorandum that directs further actions on AI and security.

7. Protect Americans' privacy, and advance equity and civil rights.

8. Stand up for consumers, patients, students, and support workers.

9. Promote innovation and competition, and advance American leadership abroad.

10. Ensure responsible and effective government use of AI.

This agenda, part of broader international discussions on AI governance, represents a significant step in the US's approach to developing safe, secure, and trustworthy AI.

Common Human Mistakes

In November 2023, a group of Australian academics apologised for including texts generated by the popular generative AI tool Google Bard.

In a letter to the Senate, Emeritus Professor James Guthrie AM – a professor in the Department of Accounting and Corporate Governance at Macquarie University – admitted to having used Google Bard to research information for a submission to a Parliamentary inquiry into the conduct of the Big 4 consulting firms, with numerous false claims being generated as reported by *The Guardian* in an article titled "Australian academics apologise for false AI-generated allegations against big four consultancy firms".

The academics' submission included case studies about alleged wrongdoing by large consulting firms that had been produced by AI but were entirely fictional.

It is believed to be the first time that a Parliamentary Committee has been forced to grapple with the use of generative AI in research and writing submissions to inquiries, which are covered by Parliamentary Privilege and free from any defamation action.

Included in the submission were a number of case studies about consulting and accounting giant Deloitte, including that it had been involved in a "NAB financial planning scandal", was sued by the liquidators of collapsed construction firm Probuild, had audited café chain Patisserie Valerie, and was auditing Westpac at the time of a scandal.

All of these claims are false.

The use of generative AI tools without fact-checking can lead to several potential issues, given the nature of AI and how it generates responses.

To mitigate these issues, it is crucial to approach generative AI as a tool for generating ideas, drafting content, or gaining preliminary insights, rather than a definitive source of information. Users should cross-reference AI-generated content with up-to-date, credible sources, especially when dealing with important or sensitive topics. Incorporating a critical evaluation process and fact-checking regimen can leverage the benefits of AI like ChatGPT while safeguarding against the spread of misinformation and maintaining the integrity of information dissemination.

When Technology Makes Mistakes

In what could be the first recorded case of wrongful arrest due to a faulty facial recognition match, a man was arrested for a crime he didn't commit. As many media outlets reported, including *The New York Times*, the arrest happened in Michigan, US.

A man named Robert Julian-Borchak Williams was in his office when he received a phone call from the police department asking him to surrender himself. He first thought that it was a prank call. That afternoon, when he drove into his home after work, a police car pulled up and blocked him from behind. The police officers got off and handcuffed him on his front lawn in front of his wife and two daughters. It was a devastating ordeal for the family of four. The police didn't say why he was arrested. They just simply showed a warrant with a photo of him on it. Police whisked him away without telling his wife where he was taken to.

The police drove him to a detention centre and took a mugshot, fingerprints, and DNA as part of the process. A day later, two detectives took him out and started interrogating him. They started questioning about his visits to an upmarket boutique shop. "When did you last go to the shop?" asked the detectives. He responded by saying that he and his wife went to see it when it first opened up years ago.

The detectives showed him the first piece of evidence, images from a surveillance camera of a man dressed in black wearing a red cap in front of a watch display. An expensive watch had been shoplifted. "Is this you?" asked the detectives.

The second piece of information was a close-up photo of the man. Robert knew it wasn't him. He picked it up, held it closer to his face, and said, "Is this me? This is not me." All black men do not look alike. He knew he hadn't committed the crime. However, he didn't know that an algorithm had wrongfully identified him. That Friday, he was about to celebrate his 42nd birthday.

In Robert's recollection, once he showed the detectives that his face didn't match the close-up on the second paper, the detectives leaned back on the chair and said, "I think the computer got it wrong."

They then handed him the third piece of paper, which he showed again that it wasn't him. He asked the detectives if he could go home. "Unfortunately not" was the answer he got.

He was kept until Friday and was released on a $1,000 bond. His family contacted attorneys in the area, and most of them requested exorbitant charges assuming he committed the crimes. An organisation interested in facial recognition technologies took up the case for him. Two weeks after his arrest, he appeared in county court. The prosecutor moved to dismiss when the case was called, but "without prejudice", meaning he could later be charged again.

A spokesperson for the prosecutor mentioned that a second witness was at the shop at the time. If this person identified him later, the prosecutor's office might decide to charge him later.

Later, the prosecutor's office informed him that he could have the case and his fingerprint data expunged. He and his wife have not talked to their neighbours about what happened. They wondered whether they needed to put their daughters into therapy. His boss advised him not to tell anyone at work. "My mother doesn't know about it. It's not something I'm proud of," Robert said. "It's humiliating."

When AI makes mistakes as happens with any automated technology, implications to humans and humanity could be significant. As professionals who are responsible for ethical practices, we have an obligation to understand both the positive and negative implications it may offer.

18

The Transformational Story of Estonia

In the heart of Northern Europe, nestled along the Baltic Sea, lies Estonia, a country whose story is a testament to innovation, resilience, and transformation. Once part of the Soviet Union, Estonia regained its independence in 1991 and faced the daunting task of rebuilding a nation from the ground up. What makes Estonia's story truly inspiring is how it embraced this challenge and turned it into an opportunity to reinvent itself as a digital powerhouse.

In the early 2000s, Estonia made a strategic decision that would set it apart on the global stage: it decided to invest heavily in digital technology and the internet. This was a bold move, considering the country's limited resources at the time, with a population under 1.4 million. However, the leaders of Estonia had a vision. They saw the potential of digital technology to transform society, government, and the economy.

One of the most significant steps Estonia took was to establish e-Estonia. This initiative aimed to digitise all government services, making them accessible online 24/7, from voting in elections to filing taxes. Estonians could now do almost everything online. This digital leap streamlined government services, reduced bureaucracy, and made life significantly easier for its citizens.

But Estonia's ambition didn't stop there. It introduced the world to the concept of e-Residency, a digital identity available to anyone around the globe. This revolutionary idea allowed digital entrepreneurs worldwide to start and run a location-independent business in the EU. Estonia became a hub for tech startups, attracting talent and investments from across the globe.

Perhaps the most inspiring aspect of Estonia's story is its focus on education. The country invested heavily in digital literacy programs, ensuring that from a young age, children were equipped with the skills needed for a digital future. Estonia's schools became some of the most technologically advanced in the world. This focus on education paid off as Estonian students consistently rank among the highest in international assessments. For example, Estonia ranked in the top ten in the UN E-Government survey in 2022 whereas Estonia's 15-year-olds are at the absolute top in Europe and in the top eight in the world in the PISA2022 survey.

The country's e-Health is another innovative solution, offering secure and efficient electronic health records (EHRs) for every citizen. Through its e-Health system, medical information such as patient histories, prescriptions, and test results are accessible to authorised healthcare providers regardless of their location. This ensures a seamless and coordinated care process, significantly reducing the chances of duplicate tests and enabling faster, more accurate diagnoses and treatment plans.

Additionally, Estonia's e-prescription service, part of the e-Health system, allows doctors to prescribe medication electronically, which patients can then pick up from any pharmacy in the country using their national ID card. The success of Estonia's e-Health initiatives is largely attributed to its robust digital infrastructure, high levels of digital literacy among its population, and a strong commitment to innovation and privacy, setting a commendable example for healthcare systems worldwide.

Estonia's journey from a small post-Soviet state to a digital leader is a powerful reminder of what can be achieved with vision, determination, and a willingness to embrace change. It stands as a shining example of how innovative thinking and a forward-looking approach can transform a nation, offering lessons in resilience and adaptation. Estonia, once a country rebuilding its identity, is now known as one of the most advanced digital societies in the world, a beacon of progress and a source of inspiration for nations striving for transformation.

Study Guide

Recommended Reading for Finance Students

This book is meticulously crafted to cater to the needs of finance professionals. It is intended to be an essential resource for students and members of professional bodies and academic institutions, aiming to bridge the gap between theoretical knowledge and practical application in the finance sector.

The table provided below offers a comprehensive overview detailing the recommended proficiency levels for finance students globally.

Chapter	Recommended Level
Introduction	All levels
Innovation	All levels
Trajectory of Technology	All levels
Artificial Intelligence	All levels
Components of the AI Ecosystem	All levels
Data	Business
Data Ingestion	Business
Data Transformation	Business
Data Labelling	Business

AI Models	Corporate
AI Model Training and Continuous Learning	Corporate
AI Model Evaluation	Corporate
AI Model Deployment	Corporate
AI Model Interface	Strategic
ML Ops	Strategic
AI Strategic Framework	Strategic
Case Study 1 – Amazing Banking Corporation & Digital Banking	Strategic
Case Study 2 – Tiny Inc. Transforms Its FP&A	Strategic
Case Study 3 – Tootal Telekom & Finance Processes	Strategic
Case Study 4 – Auroria Tax Authority Transforms the Country's Tax System	Strategic
Case Study 5 – Euro Assurance Transforms Its Global Auditing Process	Strategic
Case Study 6 – KNN Bank Overhauls Its Risk Platform	Strategic
Ethical Considerations and Legislative Landscape	All levels
The Transformational Story of Estonia	All levels
Study Guide	All levels

Approaching Case Studies

Approaching a business case study that involves leveraging AI requires a structured methodology to ensure that the technology aligns with business goals and delivers tangible benefits. Below are the key steps to consider.

1. Define the problem and objectives:

   - Clearly articulate the business problem or opportunity.

   - Set specific, measurable objectives for what you aim to achieve with AI.

2. Understand the business context:

   - Analyse the industry, market trends, and competitors.

   - Understand internal business processes, resources, and constraints.

3. Data assessment:

   - Evaluate the availability, quality, and relevance of data.

   - Identify data sources and any gaps that need to be addressed.

4. Technology exploration:

   - Research AI technologies and tools that could address the identified problem.

   - Consider the feasibility, scalability, and integration capabilities of AI solutions.

5. Stakeholder engagement:

   - Involve key stakeholders from various departments to gather insights and ensure alignment.

   - Communicate the potential impact of AI on different parts of the business.

6. Cost-benefit analysis:

   - Analyse the potential return on investment (ROI) and the costs involved in implementing AI.

- Consider both direct and indirect costs, including training and change management.

7. Solution design:

  - Design the AI solution, outlining the architecture, algorithms, and how it will integrate with existing systems.

  - Develop a prototype or proof of concept to validate the solution.

8. Risk assessment and mitigation:

  - Identify risks associated with the AI implementation, including technical, ethical, and operational risks.

  - Develop strategies to mitigate these risks.

9. Implementation plan:

  - Create a detailed plan for the deployment of the AI solution.

  - Include timelines, resource allocation, and milestones.

10. Training and change management:

  - Develop training programs for employees to understand and work with the AI solution.

  - Implement change management strategies to ensure smooth adoption.

11. Monitoring and evaluation:

  - Establish metrics and KPIs to measure the performance of the AI solution.

  - Regularly review and adjust the solution based on performance data and feedback.

12. Continuous improvement:

- Foster a culture of continuous learning and improvement.

- Stay updated on AI advancements and iterate the solution as needed.

Each business case is unique, and these steps should be adapted to fit the specific context and requirements of the case study you are dealing with.

www.ingramcontent.com/pod-product-compliance
Lightning Source LLC
Chambersburg PA
CBHW040919210326
41597CB00030B/5122